DINOSAUR
TRACKS

AND OTHER FOSSIL

FOO~~T~~

WES~~T~~ TES

DINOSAUR TRACKS

AND OTHER FOSSIL FOOTPRINTS OF THE WESTERN UNITED STATES

■

Martin Lockley and Adrian P. Hunt

artwork by Paul Koroshetz

COLUMBIA UNIVERSITY PRESS   NEW YORK

Columbia University Press
New York   Chichester, West Sussex
Copyright © 1995 Martin Lockley
All rights reserved

Library of Congress Cataloging-in-Publication Data

Lockley, M. G.
  Dinosaur tracks and other fossil footprints of the western United
States / Martin Lockley and Adrian Hunt ; artwork by Paul Koroshetz.
    p.   cm.
  Includes bibliographical references and index.
  ISBN 0–231–07926–5 (acid-free)   —   ISBN 0–231–07927–3
  1. Footprints, Fossil—West (U.S.)   I. Hunt, Adrian P.
  II. Title.
  QE845.L625   1995
  560.978—dc20   94–24198
                                                                          CIP

Casebound editions of Columbia University Press books are printed on permanent and
durable acid-free paper.

Printed in the United States of America
c 10 9 8 7 6 5 4 3 2 1
p 10 9 8 7 6 5 4 3 2 1

To

Giuseppe Leonardi and Hartmut Haubold

Leaders in the field of vertebrate track studies and pioneers in efforts to standardize and synthesize the science

and

Alwyn Williams

A pioneer in paleontology, who always inspired his students and challenged us to seek the frontiers of science

CONTENTS

FOREWORD

The study of fossil footprints has undergone a major renaissance in the last decade. The overwhelming number of dinosaur tracksites that have been found around the world in recent years has resulted in an extensive and rapidly growing number of books and publications and new sites are being discovered almost every day. *Dinosaur Tracks and Other Fossil Footprints of the Western United States* is the first book that brings together a comprehensive and illustrated review of all fossil vertebrate tracks known to date from a track-rich region that extends from Texas in the east to California in the west and from Arizona in the south to Washington in the north. The area is the best area in the world to study fossil footprints thanks to extensive exposures of Carboniferous-Recent strata in such classic localities as the Grand Canyon, Arches National Monument, and the Rocky Mountains. Many new and fascinating discoveries are reported and discussed herein for the first time and they are the result of the authors' active ongoing field research.

This book is more than just a review of the known sites in the Western United States, an area already well known for many thousands of body fossils including dinosaur skeletons. It is a demonstration of the importance of tracks for seeing vertebrate evolution in a complete new light. Many who visit the Western United States can find fossils but only a few are able to track the footprints of dinosaurs even though their trackways are abundant. Fewer still can find the tracks of pterosaurs, birds, mammals and other more enigmatic trackmakers. Even fewer have the eye for the big picture, the deductive skills and the persistence that lead to completely new and radical interpretations that can be called a revolution in the field of paleobiology. Martin Lockley and his coauthor Adrian Hunt are amongst these people.

Christian A. Meyer
Solothurn Naturmuseum
August 18, 1994

PREFACE

He who has once seen the intimate beauty of
nature can not tear himself away from it.
KONRAD LORENZ

The more I study tracks and am drawn into the fascinating world of fossil foot-prints the more I am amazed by the large number of paleontologists and geolo-gists who have devoted substantial portions of their careers to the study of rather nondescript burrows and trails of worms, molluscs, and other worthy but lowly invertebrates while neglecting the study of tracks of our vertebrate kin. Ten years ago I assumed that such neglect was due to a scarcity of footprints and to limitations in the information that footprints can provide. But the ensuing decade proved this assumption wrong.

All around the world, but particularly in western North America, trackers have been finding a superabundance of fossil footprints and learning their enor-mous utility in shedding light on a broad range of fundamental paleontological questions. With the happy combination of a naive and largely untested belief in the scientific potential of tracks and a perfect geographic location, I set out to track dinosaurs and other extinct creatures.

Looking back, I am glad I ignored those colleagues who described the study of fossil footprints as the "lunatic fringe" of paleontology. The majority will probably continue their commitment to worms and molluscs, as they desire—leaving the magnificent tracks of brontosaurs and tyrannosaurs to those of us who are truly crazy about the big beasts. Others, however, as we are beginning to see, are paying more attention to the fascinating world of fossil footprints.

The pace of discovery for trackers has picked up so rapidly that we are now overwhelmed by new data. From a personal perspective, this near-overload of opportunity has led me into a rewarding collaboration with fellow paleontologist (and coauthor) Adrian Hunt. From a general perspective, the science has come a long way in a short time. Adrian and I know we have been privileged participants in a grand scientific adventure.

Growing up in the small shires of Wales, I remember being attracted to movies of the "Wild West"—never dreaming I would come to roam the same landscapes. I arrived at the University of Colorado at Denver in 1980, the same year that Adrian Hunt also ventured from England's confined pastures to the great western desert. Here we both found the same awe-inspiring expanse of paleontologically uncharted territory. When first invited by a former student to visit a dinosaur tracksite, I embarked on the venture assuming that specialists had already interpreted the features I was about to see. Not so; we had to figure it out for ourselves.

Soon the hesitancy born of ignorance gave way to the thrill of discovery. And the lure of new data urged me on. I was becoming a tracker.

My first publications with various coauthors gave me sufficient standing in this neglected field to be regarded as an "expert," or at least as a specialist. The reports of new or long-forgotten sites began to trickle in. Site by site and season by season, my colleagues and I worked, mainly unknowingly, to rejuvenate the discipline, keeping close track of the thousands of footprints being mapped and measured in the American West. As the trickle of reports became a steady stream and then a flood, we began to think in broader terms—moving beyond a focus on individual tracks to the sites themselves and how tracksites were situated in relation to ancient landscapes. This in turn led to greater success in predicting the location of new sites. We turned away from scouring the surfaces of modern topography to tracing the ancient surfaces that had shown some promise.

In many cases the promise was more than fulfilled. More than once I have stood on an ancient surface awed by the realization that it had been trampled for miles in every direction. My colleagues and I dubbed such layers "megatracksites." One was given the special name of Dinosaur Freeway and has been considered a possible migration route.

This focus on field studies and on mapping and measuring most of the known sites has led to our discovering several hundred new sites or track layers. In this book we have attempted to describe and interpret those that we think most important. In addition to our mentors, colleagues, assistants and friends to whom we dedicate this book, we also dedicate it to future generations of trackers, who no doubt will reevaluate our efforts and greatly improve on our interpretations.

The past decade has seen a phenomenal resurgence in the study of fossil footprints, the science of *vertebrate ichnology*. Much of this progress owes to the renaissance in dinosaur paleontology. The realization that tracks are essential for understanding dinosaur behavior (speed estimates, social behavior, and so on) has done much to elevate dinosaur tracking from a trivial pursuit to a valid subdiscipline within the mainstream of geology and paleontology. And progress in the science of ichnology has generated a growing public awareness of fossil footprints.

Although dinosaurs constitute an important and often newsworthy part of the track story, dinosaurs were on the scene for only about half of the time that land tracks were being made. Vertebrate animals first developed limbs and walked on land almost 400 million years ago, in the Devonian period of the Paleozoic era. But they did not become abundant until the Carboniferous period, about 350–300 million years ago. From that time on, vertebrates left a diverse and interesting track record, now preserved in sedimentary rocks.

Vertebrate tracks can be found in rock strata in the western United States that represent more than 100 million years of earth history before dinosaurs became prominent. After the demise of the dinosaurs 65 million years ago, birds and mammals contributed a substantial and intriguing track record.

The renaissance in the study of fossil tracks has spawned an extensive and rapidly growing literature. New discoveries and interpretations are published in a wide range of technical journals, and the stories are sometimes picked up by popular science magazines and newspapers. Our own discoveries have been reported in scientific journals such as *Nature*, but they have also reached the readers of *Time* and *National Geographic*. This increased attention to tracks has helped focus scientific effort on the most productive track-bearing regions. The western United States is among the best in the world.

Few regions can match the western United States for the quality, quantity, and temporal completeness of tracksites. The eastern United States, parts of northern and southern Africa, South America, and eastern Asia do yield abun-

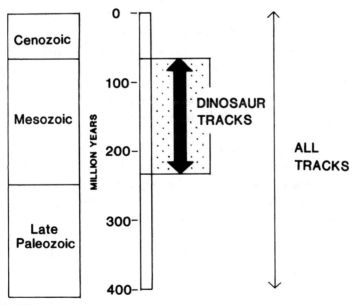

FIGURE 0.1 Vertebrates have been leaving tracks on land since the Devonian period, about 400 million years ago. Even so, much of trackmaking history falls in the Mesozoic "age of dinosaurs."

dant footprints from certain time periods. But the record in these areas is full of gaps. Europe boasts a moderately complete track record, probably because it has a long and distinguished tradition of looking for and studying fossil footprints. In Europe, tracks from all three eras (Paleozoic, Mesozoic, and Cenozoic) can be found. But the European record is patchy, and the lack of extensive outcrops and the high frequency of marine strata reduces the possibility that such an extensive track record will ever be assembled there.

By contrast, there is an embarrassment of riches in the western United States—in part because of the huge geographic area and the outcrop-rich arid terrain, but also because of the relatively complete sequence of continental or terrestrial sediments in which tracks occur. We have even coined the term "Ichnotopia" for this tracker's paradise, and we make no apologies for the unbridled enthusiasm and superlative adjectives that sometimes creep into our descriptions of some of the more spectacular sites. We are in awe of just how much new information is out there to be discovered and investigated. It is this wealth of fascinating track information that forms the subject of this book.

We undertook this study, in part, because of our obsessive interest in fossil footprints, but also because the science of vertebrate ichnology has been a neglected discipline in the fields of geology and paleontology. We also firmly believe that the study of dinosaur footprints, which now enjoys great popularity, is only a part of the overall track record. Much may be learned by applying the lessons of Mesozoic track research (dinosaur ichnology) to the study of the Paleozoic and Cenozoic track records.

The science of vertebrate ichnology is in a dynamic but healthy state of expansion, producing a landslide of new data and interpretation. Formations that were at first thought to yield only one or two tracksites often turn out to contain dozens and soon will yield hundreds. This expanding data base repeatedly forces us to restudy the track record. Sometimes known tracks and trackways, when reevaluated, are found to have been misidentified or misinterpreted. Little by little we have been learning to extract from tracks their full potential for anatomical, ecological, and behavioral interpretation. At the same time, we have been learning the importance of caution—of restraining our very human tendencies to want to read more into the data than the actual evidence can support.

In 1986 I collaborated with several colleagues to produce *Dinosaur Tracksites*. This 56-page guide to dinosaur tracksites of the Colorado Plateau and the American Southwest was published by my home institution, the University of Colorado at Denver, in conjunction with the First International Symposium on Dinosaur Tracks and Traces, which was held in Albuquerque under the auspices of the New Mexico Museum of Natural History. The guide outlined a few dozen sites, mainly in Colorado and Utah. From the moment of publication it was out of date, and it has grown increasingly inadequate. We now know of hundreds of dinosaur tracksites, in addition to hundreds revealing other types of fossil footprints.

This new and bigger book takes account of the newly acquired and burgeoning data base. And we take the opportunity to discuss the implications for existing interpretations and for the practice of vertebrate ichnology in the western

United States (and, to some degree, worldwide). Because of rapid expansion in the field, we caution the professional reader that much of what we report, especially with respect to Paleozoic and Cenozoic tracks, must be treated as preliminary. We are also aware that much of this book, too, will inevitably become outdated. Surely a number of important sites will prove to have been inadequately dealt with or even overlooked. We ourselves find or receive reports of at least one new site every week and our field excursions produce new sites every day! It is sobering to realize that somewhere on the order of fifty new sites will have been discovered between the time we write these words and the time this book becomes available.

Because we intend our book to be readable and accessible to a general audience as well as to professional paleontologists, we have tried to limit our use of unessential technical jargon. The track names are cumbersome enough, without numerous citations to often-obscure references and specimen numbers. We do, however, consider these references to be of prime importance to specialists and so have included them in the figure captions and annotated references. We have also made an effort to name those individuals (scientists or otherwise) who have been responsible for finding important tracks or track sites or for proposing important or thought-provoking interpretations. We have chosen to present the key references in an annotated form that is user friendly.

Finally, we have included an appendix that lists the museums and other major repositories of tracks and replicas in the western United States, along with a number of important tracksites accessible to the public. Our intent is not to direct visitors to sensitive sites or to encourage unauthorized collecting of specimens. Rather, we offer an educational tour and celebration of a vast, and as yet only partially charted, fossil resource in the western United States.

Martin Lockley

PREFACE TO THE REVISED
PAPERBACK EDITION

Since 1995 when this book was first published, there have been many significant new tracksite discoveries in the western United States—far too many to describe them all in detail. A number of studies have been published, however, that enhance our original interpretations. Most significant from a technical point of view is that some previously unnamed tracks now have formal names (six new ichnogenera are cited in chapters 3–5). In addition, there have been new surveys of important areas and some changes in interpretation that must be noted.

The next book in this Columbia University Press series, *Dinosaur Tracks and Other Fossil Footprints of Europe,* by myself and Christian Meyer, demonstrates many significant points of comparison with the track record of that continent. Such correlations stress that understanding the track record in one region goes a long way toward appreciating other regions and the global picture.

Martin G. Lockley
December 1998

ACKNOWLEDGMENTS

We have been fortunate in having had the opportunity to work with many other scientists and with land management officials and amateur enthusiasts who have helped us find and interpret the many tracksites discussed here. In addition to the ever-essential locality information, many of these people have contributed specimens, rubber molds, photographs, and drawings.

Many current and former members of the University of Colorado at Denver Dinosaur Trackers Research Group, which we manage, helped us in the field, and in some cases participated in the discovery and documentation of wholly new sites. Especially involved were Michael Parrish, Kelly Conrad, Gerald Forney, Rebecca Schultz, Marc Paquette, and Linda-Dale Jennings. Others who have collaborated closely with us at various times include Christian Meyer (Solothurn Naturmuseum, Switzerland); Masaki Matsukawa (Nishi Tokyo University, Japan); James Farlow (Indiana University at Fort Wayne); John Holbrook (Southeast Missouri State University); Jeff Pittman (East Texas State University); Joaquin Moratalla (Universidad Autonoma, Madrid, Spain); Vanda Faria Dos Santos and Antonio Galopim de Carvalho (Museu Nacional de Historia Natural, Lisbon, Portugal); Gerard Gierlinski (Warsaw, Poland); Jim Madsen (Dinolab, Salt Lake City, Utah); John Knoebber (Sausalito, California); and Louis Psihoyos (Boulder, Colorado).

During the past decade our fossil footprint studies have been sponsored at various times by the National Science Foundation, the National Park Service, the National Geographic Society, the Jefferson County Scientific and Cultural Facilities District, and the University of Colorado at Denver. We thank all these institutions for their support. We particularly acknowledge the efforts of Marvin Loflin, Dean of the College of Liberal Arts and Sciences at our home institution, University of Colorado at Denver. He ensured that the university could accommodate the growth of our activities by providing space and other resources.

In the course of our work, we collected hundreds of track specimens that are now mostly housed in the joint collections of the University of Colorado at Denver and the Museum of Western Colorado (CU-MWC). Additional specimens are reposited in the collection of the United States Geological Survey (USGS) at Denver. Those instrumental in establishing this curatorial infrastructure at the Museum of Western Colorado were Michael Perry and Harley Armstrong. Their work has been continued by Richard Simms, the current museum director, and Brooks Britt. We also thank Thomas Bown, Kelly Conrad, and Charles Repenning, of the USGS, in this regard. In addition we thank the Navajo Nation for prior written permission to conduct research at certain sites.

In 1991 and 1992 the Dinosaur Trackers Research Group at the University of Colorado at Denver collaborated with Masaki Matsukawa, of Nishi Tokyo University, and with the Nakasato Dinosaur Center, the Gunma Prefectural History Museum, and the Fukushima Prefectural History Museum in designing the "Tracking Dinosaurs Exhibit." This was the first paleontological exhibit exported from Colorado, and it showed to about 32,000 people each summer in Gunma and Fukushima before returning to tour Colorado in 1993 and 1994. Among the many people and institutions who helped us create this exhibit were Eddy Von Mueller, Linda Law, Rebecca Schultz, Rebecca Greben, and Kent Hups (all of the University of Colorado at Denver); David Thomas (Albuquerque, New Mexico); Hiroyuki Tanaka (Gunma Prefectural Museum); and Yojiro Taketani (Fukushima Prefectural Museum). Exhibit preparation and authorization would not have been possible without the cooperation of Teikyo Loretto Heights University in Denver, the Museum of Western Colorado, the U.S. Geological Survey, and the Colorado Historical Society. This book, especially the illustrations, has benefited from the use of ideas and materials developed during the construction of the Tracking Dinosaurs exhibit.

During the last several years we have also been involved with the growth and development of The Friends of Dinosaur Ridge, an organization established to protect dinosaur tracks and bones at historic localities near Morrison, Colorado. In 1992 the university and the Friends jointly obtained financial support from the Jefferson County Scientific and Cultural Facilities District and Amoco Production Company to undertake an extensive excavation of dinosaur tracks, which we directed. We extend hearty thanks to many of the Friends of Dinosaur Ridge, including the group's president, Bob Raynolds, vice-president Joe Tempel, and members Karen Hester, Dick Scott, Lori Marquardt, T. Caneer, and Bob O'Donnell. The Colorado Historical Society and the Colorado Department of Highways also deserve recognition for their help in that effort.

Others who have helped with access to sites and collections in the western United States and elsewhere include Dan Chure, Ann Elder, and Scott Madsen (National Park Service, Dinosaur National Monument); Don Burge and John Bird (College of Eastern Utah Prehistoric Museum, Price); Charles Pillmore (USGS,

Denver); Alden Hamblin and Sue Ann Bilbey (Utah Field House, Vernal); Kevin Padian and Tony Fiorillo (University of California Museum of Paleontology, Berkeley); Don Lofgren (Raymond Alf Museum, Claremont, California); Spencer G. Lucas (New Mexico Museum of Natural History and Science, Albuquerque); Susan Collins and Kevin Black (Colorado Historical Society, Denver); Bruce Eriksen (Minnesota Museum of Science and Technology, St. Paul); Jon Krammer (Potomac Museum Group, Minnesota); Bill Hawes and Paula Ott (Grand Junction, Colorado); Julie Howard (U.S. Bureau of Land Management, Moab); Fran and Turby Barnes (Moab); Sean Duffy (National Park Service, Arches National Monument); Jim Ferguson (Kenton, Oklahoma); Jerry MacDonald (Las Cruces, New Mexico); Camile Evans and Ted Fremd (National Park Service, John Day Fossil Beds National Monument); Debra Dandridge, David Pieper, and Jamie Kingsbury (U.S. Forest Service, La Junta, Colorado); Robert Schiller (National Park Service, Denver); John Ritenour and Clive Pinnock (National Park Service, Glen Canyon National Recreation Area); Ben Wheeler (Union County, New Mexico); Gene Foushee (Bluff, Utah); Richard Smith (Idaho Falls, Idaho); Steve and Sylvia Czerkas (Blanding, Utah); and Marc Donivan (Salt Lake City, Utah).

Finally, we wish to thank Spencer Lucas, James Farlow, and an anonymous reviewer for careful readings of an earlier version of this manuscript and for the valuable suggestions they provided. Kelly Conrad (USGS, Denver) and Heinz Kozur (Budapest, Hungary) also assisted in this regard. We also especially thank Dr Hartmut Haubold, Geiseltal Museum, Germany, for helpful discussion on a broad spectrum of track studies, and for the insight he is bringing to our collaborative study of Permian tracks. The artwork at the head of each chapter was, with one exception, produced by Paul Koroshetz, whom we thank heartily. John Sibbick kindly provided the illustration on page 1. A vote of thanks also to Ed Lugenbeel, our patient and long-suffering editor, and to his able assistant, Laura Wood. We are grateful as well for the editorial efforts of Connie Barlow and Ivon Katz.

Persons not mentioned in the acknowledgments of the first edition whom we wish to thank are John Foster, Robert Gaston, Debra Mickelson, Jennifer Schellenbach, Beth Southwell, and Emma Rainforth, who have all been associated with the University of Colorado at Denver Dinosaur Trackers Research Group in various ways. John Foster and Robert Gaston, who are both mentioned in the text, have been particularly active in making important discoveries in Colorado and Utah and making fundamental contributions to primary research.

ABBREVIATIONS OF INSTITUTIONAL
COLLECTIONS OF FOSSIL TRACKS

In some illustrations we have used the official catalogue number of the specimen. This is usually preceded by an acronym that refers to the museum or university where that specimen is housed. These acronyms are:

AMNH - American Museum of Natural History
CMNH - Carnegie Museum of Natural History
CEUF - College of Eastern Utah (Fossil)
CU-MWC - University of Colorado (Denver)/Museum of Western
    Colorado Joint Collection
LACMNH - Los Angeles County Museum of Natural History
NMMNH - New Mexico Museum of Natural History and Science
UCM - University of Colorado Museum (Boulder)
UCMP - University of California Museum of Paleontology (Berkeley)
USNM - United States National Museum

A group of six iguanodontids heading southeast on the shores of the Cretaceous Western Interior Seaway. Based on new track evidence from Dinosaur Ridge, Colorado; see chapter 5. Artwork courtesy of John Sibbick.

1

AN INTRODUCTION TO FOSSIL FOOTPRINTS

*Let us permit nature to have her way: she under-
stands her business better than we do.*
MICHEL EYQUEM DE MONTAIGNE
(*ESSAYS*, III XIII)

■ The greater Colorado Plateau region—including parts of Colorado, Utah, Arizona, and New Mexico—is a treasure trove for fossil footprints. These four states have yielded most of the ancient tracks known from the western United States, though important sites are also found in California and Texas. The paleontological record for footprints extends back to the Devonian period, almost 400 million years ago, when vertebrates (amphibians) first walked on land. However, these earliest footprints are extremely rare and are known only from a few sites around the world. No Devonian footprints have yet been found in western North America.

Footprints first became common in the Carboniferous period, sometimes referred to as the "age of coal swamps" or the "age of amphibians." It was at that time, about 300 million years ago, that the western track record began. The trail of fossil footprints continues into the Permian, which is the final period of the Paleozoic era, then into the Mesozoic era, the "age of reptiles," and finally into the "age of mammals" of our own Cenozoic era.

This long and magnificent record of fossil footprints in the West owes in great measure to fortunate geological circumstances. Some parts of the greater Colorado Plateau region (figure 1.1) offer a nearly complete sequence of terrestrial sedimentary deposits from Carboniferous times through to the present. Only for a brief period, during the late Cretaceous, did the shallow sea that inundated what are now the plains states and provinces rise enough to flood extensive regions of the western uplands, thus preventing the accumulation of a track record of land-based animals. With the exception of this one widespread marine interlude, the seas only encroached locally onto the margins of western North America throughout the past 300 million years, thus leaving a predominantly terrestrial sequence of sediments that contains one of the best records of land animals found anywhere. The Mesozoic record of dinosaurs and the Cenozoic record of mammals in the region of the Colorado Plateau are particularly famous.

Another reason the track record of the western states is so good is simply that footprints are easy to see. Track-bearing rocks are very well exposed, as much of this area is semi-arid desert with only sparse vegetative cover. The rock exposure in canyonlands country is especially good—a veritable geologic paradise, and excellent tracking terrain. On average, a new track site is discovered in the West each week (figure 1.2).

In this book we use fossil "tracks" and "footprints" interchangeably. Both terms refer to what geologists call trace fossils—in contrast to body fossils, which (for vertebrates) are the actual skeletal remains. Trace fossils can be used to understand some aspects of the anatomy of the organism that made it, but they are particularly useful for interpreting behavior. A sequence of consecutive tracks is called a "trackway"—sometimes also, a "trail." A trackway sheds light on the posture, gait, and speed of the trackmaker. However, there is more to the study of trace fossils than a simple analysis of locomotion and behavior.

The study of trace fossils is the field of ichnology, from the Greek word *ichnos* ("trace"). In addition to offering clues about the behavior of individual animals, trace fossils can provide insights into social or group behavior and the makeup and distribution of ancient animal communities. For example, tracks may show approximately how many individuals of a particular kind of animal were present in a given area and their size range. Tracks can also indicate which species shared the same habitat and how they might have interacted. Similarly, track types help us identify different animal groups and determine which types were abundant. And by studying the type and features of the rock in which the tracks occur, we can learn more about the habitat and ecology of the extinct animals. This is the science of ancient ecology, or paleoecology.

In addition to ichnology and paleoecology, fossil tracks can be used to ascertain paleobiogeography. Data from several dozen tracksites within deposits of the same age can give a regional picture of the geographic extent of an animal's range. Tracks and tracksites from particular time periods may be compared with those from other regions in North America and around the world. In this way tracks may also prove useful in accurately assessing the age of rock units. This application of tracks to deciphering the age of rocks is similar to the application of other fossils in chronological studies. This is the science of biostratigraphy or palichnostratigraphy (meaning, ancient track stratigraphy).

Subtle or sudden changes in the composition of track assemblages may appear in successive layers of track-bearing strata. These may relate to changes in the ancient environment that in turn affected the ecology; or the changes may tell us something of the patterns of evolution and extinction that affected individual species or groups of track-makers. Abrupt changes may also indicate gaps in the fossil record—places where substantial sedimentary sequences are missing because of a hiatus in sediment deposition or because of later erosion that carried

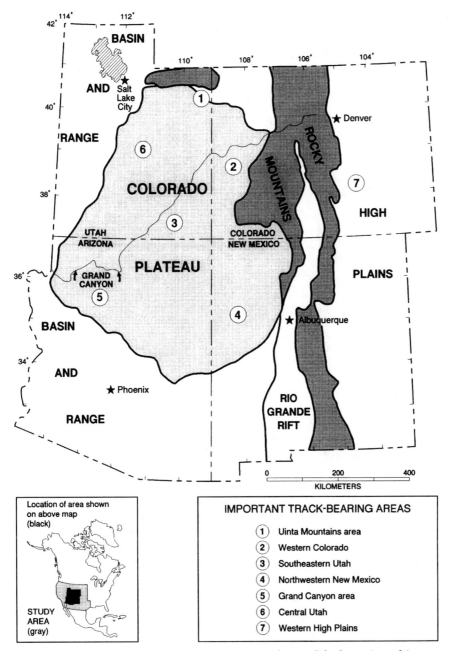

FIGURE 1.1 Map of Colorado Plateau region, with simplified stratigraphic sections.

(continued)

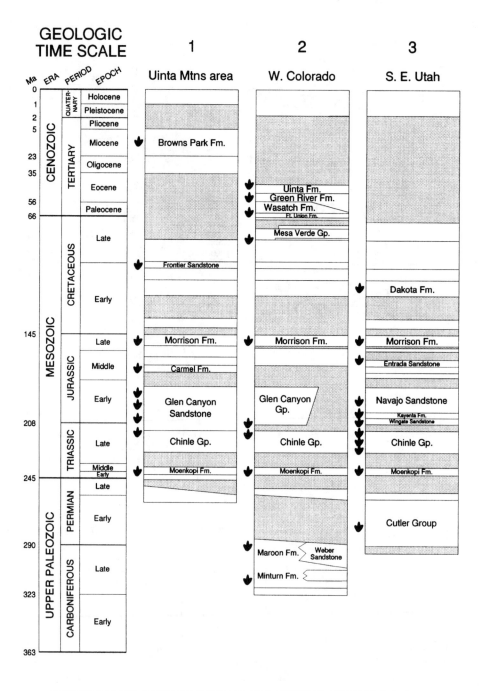

GEOLOGIC TIME SCALE

Ma	ERA	PERIOD	EPOCH
0	CENOZOIC	QUATERNARY	Holocene
1			Pleistocene
2			Pliocene
5		TERTIARY	Pliocene
			Miocene
23			Oligocene
35			Eocene
56			Paleocene
66			
	MESOZOIC	CRETACEOUS	Late
			Early
145		JURASSIC	Late
			Middle
			Early
208		TRIASSIC	Late
			Middle / Early
245	UPPER PALEOZOIC	PERMIAN	Late
			Early
290		CARBONIFEROUS	Late
323			Early
363			

1 — Uinta Mtns area

Browns Park Fm.

Frontier Sandstone

Morrison Fm.

Carmel Fm.

Glen Canyon Sandstone

Chinle Gp.

Moenkopi Fm.

2 — W. Colorado

Uinta Fm.
Green River Fm.
Wasatch Fm.
Ft. Union Fm.

Mesa Verde Gp.

Morrison Fm.

Glen Canyon Gp.

Chinle Gp.

Moenkopi Fm.

Maroon Fm. / Weber Sandstone

Minturn Fm.

3 — S. E. Utah

Dakota Fm.

Morrison Fm.

Entrada Sandstone

Navajo Sandstone

Kayenta Fm.
Wingate Sandstone

Chinle Gp.

Moenkopi Fm.

Cutler Group

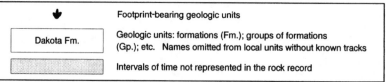

Footprint-bearing geologic units

| Dakota Fm. | Geologic units: formations (Fm.); groups of formations (Gp.); etc. Names omitted from local units without known tracks |

Intervals of time not represented in the rock record

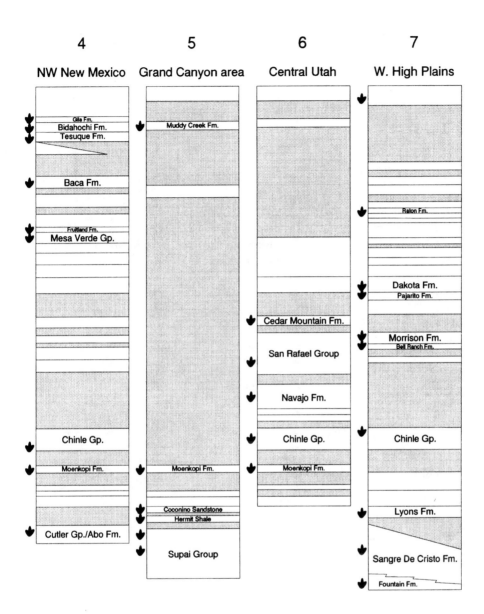

FIGURE 1.1 (continued)

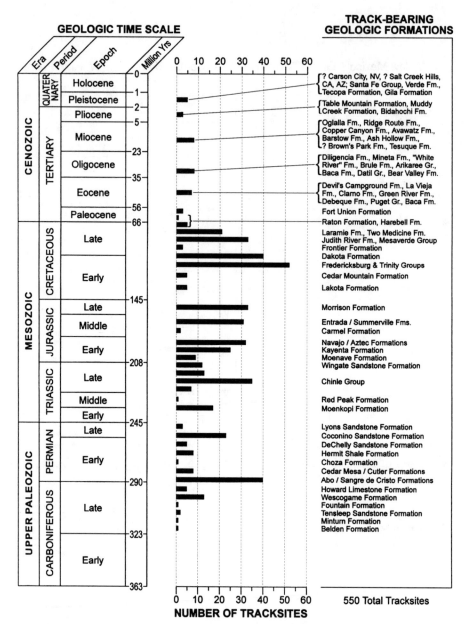

GEOLOGIC TIME SCALE

TRACK-BEARING GEOLOGIC FORMATIONS

NUMBER OF TRACKSITES

550 Total Tracksites

FIGURE 1.2 Histogram of number of fossil footprint sites in the western United States.

away previously deposited sediments. In such instances older track-bearing rocks are overlain by much younger track-bearing strata.

Tracks evoke a peculiar fascination in this respect because they represent animals that were active at times when sediments were not accumulating. Most bones that become fossils are buried while sediment is being deposited, but tracks mainly represent a record of animal activity at other times. It is the job of paleontologists to combine a knowledge of biology and geology to decipher the signals left in the fossil record and to distinguish patterns of track distribution that result from biological causes (ecology, evolution, and extinction) from those relating to geological causes (sedimentology, stratigraphy, and preservational factors).

How to Observe and Record Tracks and Trackways

Tracks vary considerably in size and shape, and trackways vary in the number and distribution of prints. A systematic approach for describing and recording them is essential.

The first step in measuring vertebrate tracks and trackways is to record the size and shape of individual hind and fore footprints and the distances and angles between steps (figure 1.3). The hind foot is also known as the *pes* and the fore foot as the *manus*. Measurements of individual footprints tell us how big the trackmaker was, how many weight-bearing digits it had on its feet, and whether it had larger hind feet than fore feet, as in most dinosaurs, or larger front feet than hind feet, as in many mammals. Using our own hands and feet as a model, we can count toes or fingers one to five (I–V), following the convention that the thumb and big toe are I and the little toe and finger are V.

Animals whose limbs end in five digits, and tracks with five-digit impressions, are referred to as *pentadactyl*. Humans, bears, and some lizards—and the tracks that they make—are prominent examples today. *Tetradactyl* trackmakers

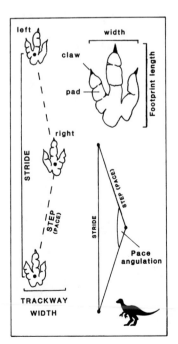
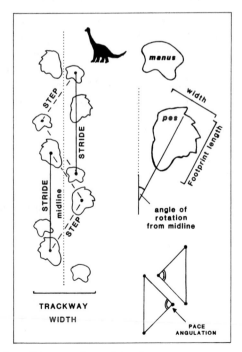

FIGURE 1.3 Measurements used in describing tracks and trackways. *After Lockley 1991.*

have four functional digits on each limb, as in hippos, cats, dogs, and some sala-manders. *Tridactyl* trackmakers, such as tapirs and some birds, are three-toed. A two-digit or functionally *didactyl* condition is common today in cud-chewing mammals (ruminants), such as cattle, sheep, and deer. These cloven-hoofed mammals are known as "even-toed ungulates" (ungulate means hoofed). A single-toed *monodactyl* condition characterizes horses within the living "odd-toed ungulates."

Counting the number of toes in a fossil print is the first step in its identification. As shown in figure 1.4, there are many fossil examples of tracks with different numbers of toe impressions. It is these fundamental differences in shape and size of tracks, as well as the configuration of the trackways, that ultimately allow us to identify the trackmakers. We must exercise caution, however, in noting that not all of an animal's toes may touch the ground. For example, four-toed birds often leave tracks with three-digit impressions.

In any discovery of footprints it is important to record the trackway configuration, which includes step and stride length, trackway width, and the degree to which feet are placed at an angle to the general direction of travel. This information will help determine, first, if the animal was two footed (bipedal) or four footed (quadrupedal). Did it have long legs and a short body, like a camel, or short legs and a long body, like a hippo? With four-footed animals we can reconstruct the trunk length by using the trackway to estimate the positions of the hip and shoulder sockets (figure 1.6). Trackway data also reveal whether an animal was an erect walker, with a narrow trackway, or a sprawler, with a wide trackway. We will also learn if the trackmaker's feet rotated in (pigeon-toed) or out (duck-footed). Finally, we can determine if the animal was walking, running, hopping, or moving in some other fashion.

Fossil tracksites generally reveal footprints of more than one animal and often several different species or types. At sites with dozens—even hundreds—of trackways, we can make a map of the whole site and count the different individuals that made the tracks and from this derive the relative abundance of each species or type (figure 1.7).

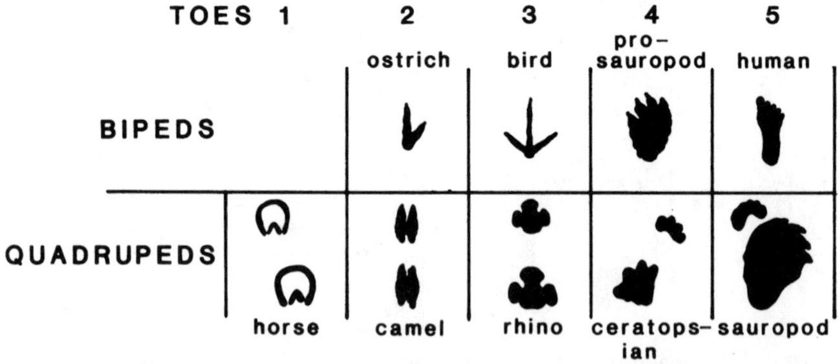

FIGURE 1.4 Diversity in toe number in bipeds and quadrupeds.

Who Dun It?

One of the hardest questions to answer in the study of fossil footprints is Who dun it? In the study of modern tracks it is always possible to identify the trackmaker, because sooner or later one can see the animal making footprints. But with fossil tracks one can never identify the species of trackmaker with absolute certainty. This does not, however, prevent the observer from drawing educated conclusions about which animals were the likely trackmakers.

Using existing anatomical knowledge about the different morphologies of the feet of various groups of extinct animals, it is easy to distinguish fossil tracks among the broadest groups of organisms. For example, brontosaurs and birds have quite distinctive feet, and within such large natural groupings as the birds, even in the fossil record, it is easy to differentiate among the tracks of water birds—such as the web-footed ducks, the large herons, and small waders. Similarly, anybody can distinguish the tracks of fossil carnivores from those of hoofed ungulates or elephants, and it requires only elementary tracking skills to differentiate more specific groups. With a little knowledge one can learn to discriminate between the tracks of dogs and those of cats (both carnivores) and between deer and camels (both ungulates).

The distinctive shapes of the feet of various major animal groups thus allows us to approach the study of fossil footprints with some confidence. Even though some fossil footprints have been attributed initially to the wrong groups of trackmakers, there is general consensus about which groups of animals were responsible for many distinctive footprint types.

It is always desirable to match tracks with foot skeletons when undertaking a difficult identification. For example, the tracks of horned dinosaurs (ceratopsids) are surprisingly rare (see chapter 5),

(continued)

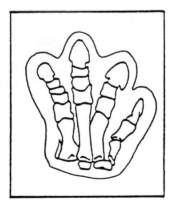

FIGURE 1.5 Comparison between ceratopsian foot skeleton and purported ceratopsian track shows a very good correlation. All tracks that have been "identified" can be checked this way if foot skeletons are available.

despite the fact that their skeletal remains are common. When purported ceratopsid tracks were first discovered, quite recently, it proved helpful to compare the foot skeleton with the large tracks, which did demonstrate a good match. Another key piece of evidence was that the ceratopsid tracks were discovered in strata of exactly the age from which ceratopsid bones are also known. In fact, a good case can be made that ceratopsids were the only group of animals capable of making such tracks at that time.

In cases where a track cannot be matched with a foot skeleton of the right age, care must be taken in attempting to interpret the trackmaker. The discoverers should not give the tracks names that imply that the trackmaker has been positively identified. Examples of groups where the "who dun it" question is hard to answer include various archosaurs, which is the broad subgroup of reptiles that includes crocodiles and pterosaurs as well as dinosaurs. Among the archosaurs of the Mesozoic, the bipedal carnivorous dinosaurs (theropods, like *Allosaurus*) and the bipedal herbivorous dinosaurs (ornithopods, like *Iguanodon*) have quite similar feet. This problem of distinguishing footprints has always plagued ichnologists, and not just those concerned Mesozoic dinosaurs.

A good approach to inferring a trackmaker at any particular time is to compile a list of suitable animals that existed during that age or epoch, and then to seek the best possible correlations. In addition, the feet of such animals, if they are known from fossil remains, should be carefully measured and compared with the footprints. If the bones with the closest anatomical fit to a particular track are found only in strata of a different age, however, the similarity may be misleading. It may not imply that the same species or even genus existed for the millions of years before or after the time indicated by the bones alone.

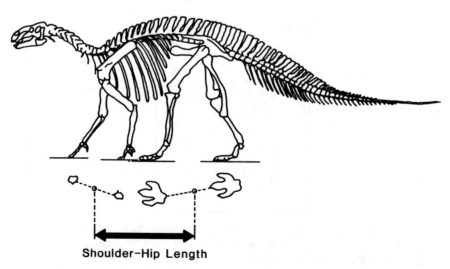

Shoulder–Hip Length

FIGURE 1.6 Estimating body, or trunk, length (from shoulder to hip) from trackways of four-footed animals.

The site map is an invaluable tool for trackers. By laying out a simple grid, with compass and tape, string or a chalk line, it is easy to plot the position of all trackways and individual tracks. The site map (figure 1.7) then serves as a basis for distinguishing and locating each individual trackway, and, if necessary, assigning them numbers for measuring and record-keeping purposes. The site map also clearly shows the direction of travel of each individual trackmaker. These data can also be plotted on an orientation diagram or a "rose diagram" (so-called because of its flower-like appearance). Such diagrams are easily plotted simply by recording how many trackways are oriented in each direction. Typically we divide up the familiar 360° compass field into convenient sectors of 15⁻. This is demonstrated for the Sloan Canyon tracksite map which records 26 Late Triassic trackways from northeastern New Mexico with the corresponding rose diagram and census (figure 1.7).

As shown in this Triassic example, it is easy to derive a census of the number of different trackway types. This gives us an indication of the proportion of different trackmakers that were active at the site. Another way of counting or assessing the abundance of particular trackmakers is to record how often each type occurs at various tracksites (figure 1.7 and 1.8). An animal may appear to have been common at one site but rare or unknown at all other study sites. Thus it may have had a limited geographic range.

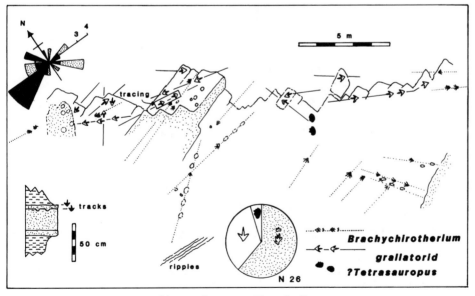

Sloan Canyon Tracksite

FIGURE 1.7 Map of the main Sloan Canyon tracksite, showing the relative number of different types of trackways. Rose diagram (top left) shows preferred westerly orientation of *Brachychirotherium* trackways (black) and the variable orientation of tridactyl trackways (stippled).
After Lockley and Hunt 1993 (see chapter 3 for details).

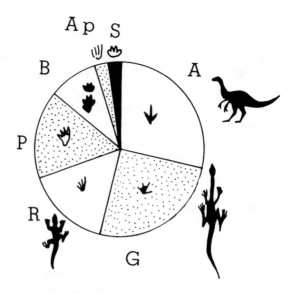

CENSUS BASED ON TOTAL TRACKWAYS   N : 32

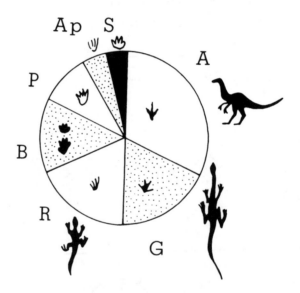

CENSUS BASED ON   LOCALITIES N : 21

FIGURE 1.8 Proportions of fossil track types provide a census of ancient animal communities. This example is drawn from late Triassic tracks data from the Dinosaur National Monument area. A = *Agialopous*. G = *Gwyneddichnium*. R = *Rhynchosauroides*. P = *Pseudotetrasauropus*. B = *Brachychirotherium*. Ap = *Apatopus*. S = (unspecified) synapsid. (Compare with Figure 3.23.)

Trackway censuses may also help us to determine the ratio of predator to prey species. Track sites from the Triassic to the Pleistocene have proved useful in this regard. They often, though not always, provide data on types of trackmakers that are reasonably consistent with the body fossil record.

Track censuses reflect many of the major faunal changes that have taken place between successive ages, epochs, and periods—often adding information that helps refine our understanding of these faunal shifts. If, as is often the case, we find that the track record is consistent with patterns of evolution determined from bones, then we can consider these tracks reliable and look to them for additional paleontological information. If, on the other hand, the track record is not consistent with data derived from skeletal remains, then we need to ask why. Is it because the track record is biased toward overrepresentation of animals that were more active or more inclined to frequent wet areas where they could make tracks? Or is the body fossil record in fact biased and misleading? These are intriguing and challenging questions, which we shall return to in chapter 7.

Usually, the best way to answer such questions is to find more data. This is the way science typically works, especially in the study of vertebrate tracks. Only recently have the data expanded to the point that tracks can be used with confidence in drawing paleoecological or paleobiogeographic conclusions. For example, the census of trackways from the late Triassic of Dinosaur National Monument, shown in figure 1.8, was compiled by us in 1991. We made further discoveries of tracks there in 1992, and the data base increased considerably. This allowed us to compile a new census, which we will discuss in chapter 3. The 1992 data do not radically alter the assessments of the relative abundances of different species, but the new information does add to the list of trackmakers and alter certain relative abundance estimates.

An ever-improving track record is thus no less important than an ever-improving bone record, archaeological record, or tally of stars. In all the sciences new field discoveries and measurements are the essential raw materials that theorists need to formulate and test hypotheses. The accumulation of new track data has, in fact, allowed us to demonstrate how frequently particular track types or combinations of track types occur in strata of particular time periods or particular ancient environments.

Naming Tracks

Like living animals and fossil bones, fossil footprints and other trace fossils are given scientific names. Scientific names are sometimes tellingly descriptive. A dog becomes *Canis familiaris* (a familiar or domesticated canine); *Homo sapiens* are the "same," and supposedly "wise," creatures we see when we confront our reflection in a mirror. Although the names are latinized and not always user friendly, they do have the advantage of consistency.

The naming system in use today was introduced by the Swedish botanist Carolus Linnaeus (1707–1778). This Linnaean system is also known as the system of binomial classification because of its double-barreled format: genus name first and species name second. All living plants, animals, and fossils, once discovered or known, must be given scientifically acceptable genus and species names, accompanied by standardized descriptive documentation published in a recognized scientific journal.

Fossil vertebrates like *Tyrannosaurus rex* (the king of tyrant lizards) are named in just the same way as living species. Fossil names are often based on diagnostic features of appearance, the place of discovery, the discoverer, or some well-known person related to that field of research (whom the discoverer wishes to honor). For example, the name *Mylodon darwinii* was given to an extinct species of ground sloth.

The same Linnaean system of binomial classification that is used for living and fossil animals and plants has also been applied to trace fossils. This tradition began in confusion when names were applied to invertebrate traces that had been mistaken for fossil algae. Despite this inauspicious start and the fact that some scientists objected to the use of the Linnaean system for naming traces of activity rather than actual organisms, the tradition has survived and has been widely used.

Like most systems that stand the test of time, the formal naming of trace fossils has merit. The system is now well established and ingrained in the scientific literature. But more than this, it is designed to provide a clear distinction between body fossils and trace fossils. The naming of trace fossils, or ichnotaxonomy (meaning, trace fossil taxonomy), is technically considered a parataxonomy (meaning, "alongside" taxonomy) because the name is not derived from and equated to a known body fossil. The two taxonomies, rather, are independent, though obviously parallel.

An example of another parataxonomy is the Linnaean system of binomial classification applied to fossil pollen and spores. Fossil pollen and spores are not always found in intimate association with leaves and branches, and thus the plants that produced them are not always identifiable. Nevertheless, the utilization of fossil spores and pollen in geologic studies is very important. Names are essential. A particular fossil pollen or spore thus may have one name and the plant that produces it another. The same holds for fossil footprints and the animals that made them.

The importance of scientific names is demonstrated by the fact that there is an International Code of Zoological Nomenclature (ICZN) that governs naming procedures. Scientists associated with the ICZN have recently reviewed the use of the Linnaean system in naming trace fossils and concluded that ichnotaxonomy is a valid scientific practice. Ichnologists may thus continue to refer to ichnospecies and ichnogenera in the same way that biologists refer to species and genera.

Many vertebrate paleontologists may suspect or even believe that, for example, *Tyrannosauripus* is the track made by *Tyrannosaurus*. But until exceptionally sound evidence is provided (as discussed in chapter 5), the correlation is not assured. Ideally, the animal must be found, quite literally, at the end of its trail. One famous example is of a Mesozoic horseshoe crab found dead in its tracks (figure 1.9). Even so, the trail has been given a distinct name.

More typically, fossil tracks and traces represent animals that can only be identified in a general way. For example, a reference to "sauropod" (or "brontosaur") tracks does not identify a particular species or even genus of track-

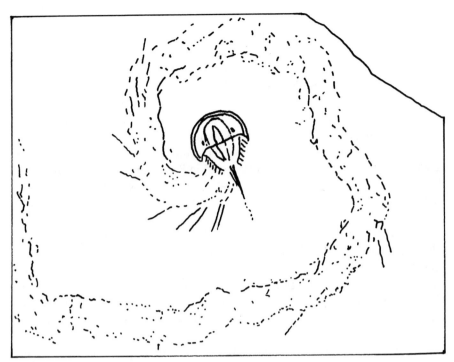

FIGURE 1.9 A Mesozoic horseshoe crab found dead in its tracks.
After Barthel, Swinburne, and Conway Morris 1990.

maker. Rather, such general categories typically identify the trackmakers at the broader level of family or order.

There are also many examples of tracks that were made by trackmakers whose taxonomic affinities are completely unknown. In some cases the trackmaker's skeletal remains have never been preserved or discovered; in others, paleontologists have not yet been able to make a correlation between the tracks and a convincing trackmaker candidate. For these reasons it is easy to see the need for clear distinctions between the names of organisms and the names of their traces.

Although there is some validity in the criticism that footprint names are cumbersome and confusing because they clutter the literature and provide a whole new array of names, they are as essential to the ichnologist as Latin binomials are to the biologist. Scientific names for tracks are necessary if we are to discriminate among a wide variety of track types that have no common names. Many trackers have labeled footprint names by ending them in "pod," "podus," "pes," or "ichnus" (all of which mean foot or foot trace). For example, *Brontopodus*, *Tyrannosauripus*, and *Brasilichnium* respectively indicate "brontosaur footprint," "tyrannosaur footprint," and "trace from Brazil." In principle, this is the same as referring to a dog track or a raccoon track. The only difference is that modern tracks have never been assigned a distinct set of latinized scientific names, whereas most fossil footprints are named formally.

In this book we use track names wherever they are necessary, but we avoid using names that are obscure or problematic in some way. This is not to say that we shirk our duty to explain complex issues where important scientific principles are at stake, but we believe that the nomenclatural tools of ichnology should be used with caution. We also try to avoid excessive detail by using the genus or ichnogenus name alone, rather than always using the double-barreled alternatives. We do, however, explain the names' derivations wherever possible and discuss the case histories of important track names.

Fossil tracks were first named in the nineteenth century, when footprints from Scotland that were of Permian age (245 to 285 million years old) were described in the literature. In 1836 the first dinosaur tracks (found in Connecticut) were named by Edward Hitchcock. The name he chose was *Ornithichnites* (meaning, bird or bird-like traces).

In most cases trackers have followed the conventions employed by those who work with actual skeletal remains; they have used the binomial classification. The only difference is that the genus and species names are referred to as ichnogenus and ichnospecies names to indicate that they refer to trace fossils. Hitchcock himself chose to use the binomial system to name one of the largest three-toed footprints discovered in those early days of fossil hunting: *Eubrontes giganteus* ("large true thunder").

Some workers have taken shortcuts by referring to *Iguanodon* tracks—thus using the name of the inferred trackmaker. It is tempting to do this when distinctive tracks are found in the same deposits as skeletal remains of good candidates for potential trackmakers, as was the case with *Iguanodon*. However, a safer approach is to refer to tracks of an *Iguanodon*-like animal. Accordingly, if a track has not already been assigned to an ichnogenus, we adopt the cautious approach of using generalized trackmaker categories, such as tracks of carnivorous dinosaurs, or duckbills, or brontosaurs.

How Are Tracks Preserved?

Unlike the relatively robust bones of large animals, footprints can easily be destroyed by the first rain or wave that washes over them. They may also be obliterated by the next herd of animals that passes through the area. The growth of vegetation may work its damage more slowly, but just as effectively. All in all, the likelihood that any single track or trackway will become a fossil is exceedingly small.

On the other hand, every tetrapod animal can potentially make hundreds of thousands, even millions, of tracks in its lifetime, whereas it only has one skeleton consisting of a few hundred bones. In theory, if an animal were to leave just one track every week that made it into the fossil record, that individual would have contributed over a thousand footprints in the course of a twenty-year lifespan. Thus the first factor to consider in accounting for the relatively common occurrence of preserved fossil footprints is simply that track production rates far exceed the production of skeletal remains in any population or species (figure 1.10). Moreover predators, scavengers, and decomposers do not attack tracks in the way that they consume carcasses.

Studies of modern tracks reveal that footprints nevertheless deteriorate rapidly; they are usually destroyed within a few days or weeks. This has led to

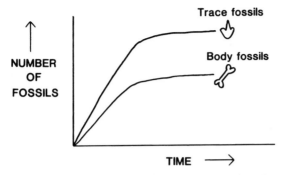

FIGURE 1.10 Why fossil footprints are more common than bones. Note that the number of tracks that a species can make in a given period of time (upper curve) is always potentially more than the number of bones that it can leave in the body fossil record (lower curve).

the popular "cover up" explanation of track preservation. In this scenario, tracks made on a wet surface, such as a mudflat, are baked by the sun until they become rock hard. The tracks are now resistant to erosion by the next tide or flood, which may also cover them with a new layer of sediment as the water subsides. We have little doubt that this type of explanation is valid in many cases. For example, rock surfaces that bear fossil tracks often display mudcracks, which proves that the surface was wet at first but then dried out.

There are, however, other explanations. One we favor pertains to the phenomenon of underprints. An underprint, also known as a "ghost print," represents the impression of a foot transmitted to a layer, an underlayer, beneath the actual sediment surface (see figure 1.11). By their very nature, underprints are already buried at the time they are made. This means that there is no problem explaining how they can be covered by the next layer of sediment without being eroded away. Many fossil tracks are found in sediments that show all the signs of having accumulated in very wet, swampy settings. In such sediments there is no evidence of mudcracks or the desiccation of substrates to produce rock-hard, erosion-resistant surfaces. On the contrary, the surfaces remained wet most of the time and the animals, particularly large ones, sank in very deep, leaving underprints at depths less vulnerable to perturbation.

A little reflection on the phenomenon of trackmaking leads us to another consideration about the dynamics of track preservation. Most tracks with any appreciable depth are protected from the full force of erosion. And because a track is a sunken indentation it will accumulate additional sediment more readily than will the surrounding surface. Tracks are thus excellent sediment traps. A fine example of this can be seen at Dinosaur State Park in Connecticut, where a set of large Jurassic tracks named *Otozoum* (meaning, giant animal) are on display. The tracks clearly show skin impressions, thus proving that they are true tracks and not underprints; yet the area in between and around the tracks is scoured by erosion. This attests to the considerable erosive power of the sand-laden currents that swept in and buried the tracks, but proves that the tracks themselves were protected by their very shape and depth. As depressions or indentations, they were sheltered, low-energy pockets—immune to the scour of sand and water just centimeters above. They therefore acted as miniature sedimentary basins or traps.

Where Are Tracks Preserved?

Successful tracking depends on knowing where to look. In this section we discuss the local "where" of rock type rather than the broader geographic "where" of paleontological provinces. The best way to narrow one's quest is to begin with first principles and to adopt a process of elimination. One proceeds by deciding where it makes sense *not* to look.

Obviously, most vertebrate tracks are made on land. Thus we must begin by looking in continental, or terrestrial, sedimentary deposits, not in marine sediments. But the best rock prospects are usually those that represent habitats that are neither completely dry nor deeply submerged. Wet substrates adjacent to bodies of water (fresh, salt, or brackish) are ideal for track impressions and their ultimate preservation. Shoreline deposits are thus a prime target. Indeed, the history of discoveries confirms that tracks are most common in deposits that accumulated along lake shorelines, in wetlands and swamps associated with coastal plains, in floodplain environments adjacent to rivers, and around ephemeral or playa lakes in arid or desert environments.

It may not be easy for the nongeologist to recognize these ancient environments among the many sedimentary rock types that exist, so a simpler approach can be adopted. Unlike body fossils, such as bones and shells, which usually occur *within* a bed or layer, tracks occur *on* the surfaces (bedding planes) between strata (figure 1.12). This is because the vast majority of footprints are made in between episodes of sedimentation. They represent the activity of animals during periods when the accumulation of sediments has been interrupted. These breaks in sedimentation, sometimes referred to as hiatuses, are common and they account for the differences, and bedding planes, between sedimentary layers in rock successions. The breaks may represent only a few hours or days, as in the case of sediment accumulations controlled by tidal cycles or run-off in river valleys. In other cases the breaks may represent years, as in the case of ephemeral lakes. Some breaks represent much longer periods, on the order of millions of

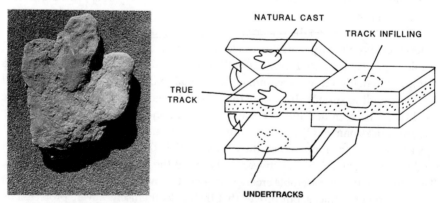

FIGURE 1.11 Tracks may be preserved as true tracks, naturals casts, undertracks, or track infills. Photograph shows an example of a natural cast of a Jurassic theropod track from eastern Utah.

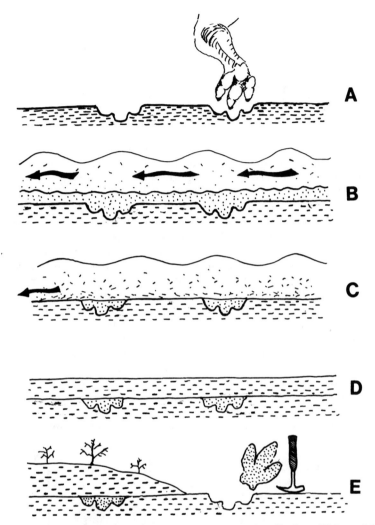

FIGURE 1.12 Tracks make good sediment traps and will often fill in quickly, before the area between the tracks is covered over.

years, and are recognized by geologists as major "unconformities." An unconformity marks a significant slice of the sedimentary record that is outright missing in a particular area—often as a result of a long-term rise or fall in sea level, during which time the active shore environment has been shifted to another location.

The break in sediment accumulation signaled by an unconformity allows for the possibility of trackmakers passing by and for tracks to have been preserved on the surface that immediately underlies the unconformity. In principle also, the longer the break, the more chance there is that trackmakers would have ventured into the area. Of course, the probability that tracks will be preserved depends on the condition of the substrate and other factors just discussed. Nevertheless, we can use these basic principles to conclude that *the best place to look for tracks is on bedding planes that represent surfaces that were exposed between episodes of sedimentation.* This strategy for discovering tracks does, in fact, work extremely well: 99

Elite Tracks

Are some tracks more important or better than others? The answer is Yes! What makes one track or trackway better than another, in the vast majority of cases, is that it is better preserved. Good preservation is important for a number of reasons, but mainly to ensure that trackers derive accurate information on foot shape and other trackway parameters. In fact, several trackers have stressed that new track types should only be named in cases for which adequate supplies of well-preserved material are available for study and for which accurate measurements can be taken. If tracks are poorly preserved, they may provide misleading information on foot size and shape, step lengths, and trackway widths that in turn may lead to dubious interpretations.

Ichnologists have also noted this problem of differential quality of preservation in the study of invertebrate trace fossils. In many cases several generations of burrows are superimposed one on another in a single layer of strata. Some, usually the older burrows, are very indistinct, whereas others, presumably those made most recently, are clear. Clear burrows that are superimposed on older traces, or traces that have been better preserved in some way, are called "elite trace fossils." Exactly the same thing has been observed in vertebrate paleontology in the study of fossil footprints. Particularly on surfaces that have been extensively trampled, the first-formed (oldest) tracks are often obscure because they have been damaged by exposure to more trampling and more weathering. By contrast, the most recently formed (youngest) tracks on the same surface will stand out clearly. Such footprints can be called "elite tracks."

(continued)

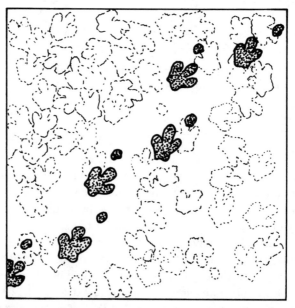

"Elite tracks" refer to certain types of well-preserved tracks that are found superimposed on poorly preserved tracks.

Trackers should not be elitist, however. We must not ignore tracks that are poorly preserved. They may not be identifiable today, but experience may help us to identify them in the future. And even poorly preserved tracks may offer some new information, ecological if not taxonomic, to the overall picture. Poorly preserved tracks may also be very useful in sedimentological studies of ancient substrate conditions. However, when it comes to naming tracks and conducting detailed biological analyses, elite tracks are clearly preferable.

percent of all known tracks are associated with bedding plane surfaces. Trackers should also note that the best hunting is done when these surfaces are illuminated by low-angle light. Bad, overhead light can make tracks very hard to see and cause us to overlook many subtle traces.

Ichnologists now recognize that many sedimentary units are capped by track-bearing layers. Each unit represents a cycle of deposition, such as a flood or a season of sediment accumulation, that ended with exposure of the depositional surface (figure 1.13). This can be seen at many scales—from localized crevasse splay sediments, which represent a river breaching its bank, to track-bearing layers that cover hundreds or thousands of square miles. The latter we have termed "megatracksites." These, too, can be explained in terms of cycles of sediment accumulation and nondeposition (unconformities) that alternately cover and expose large areas as sea level fluctuates. As described later, there are also some interesting new observations emerging about why certain formations contain more tracks than others.

Thus we conclude that the answers to *where* and *how* tracks are preserved are largely geological, not biological. Obviously it is necessary for animals to pass by (a biological activity) in order to leave tracks. But the substrate conditions must be suitable in the first place, and then the dynamics of sediment accumulation must be conducive to preservation.

Interpreting Fossil Footprints

A rather disturbing current of sloppy and unscientific speculation has marred the fossil footprint literature. Overenthusiastic speculation arises from the natural tendency of observers to want to interpret footprints in terms of behavior. While it is true that fossil tracks were made by once-living animals, these animals are now extinct and we can no longer observe them. Moreover, the tracks have been modified in various ways between the time they were made and the time they were preserved in the geological record. It is thus essential to keep a geological perspective in mind when interpreting fossil footprints.

The geologic process of track modification is the subject of the field known as track taphonomy. The science of taphonomy in general, dealing with the postmortem to final burial history of a fossil, has become an established subdiscipline in paleontology in the last two decades. By contrast, track taphonomy is a com-

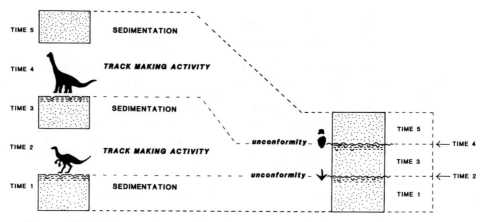

FIGURE 1.13 The occurrence of tracks at the exact interfaces between sedimentary layers (beds, members, formations) signifies trackmaking activity between episodes of sedimentation. Tracks also often occur in association with unconformities, in which units of strata are missing.

paratively new concept. Although it involves similar principles (i.e., period from the time of trackmaking to time of track burial), track taphonomy also gives rise to some unique considerations.

For example, except in the rare case of movement of an entire block of strata, a track cannot be washed away to another location or a much-later time horizon, as can a single bone from a skeleton. And tracks can produce undertracks, which have no equivalent in the body fossil record. On the downside, it may be hard to distinguish tracks that were poorly preserved at the time that they were made from those that initially were well preserved but then deteriorated during the period of taphonomic modification. With these three considerations in mind, we offer the following guidance: Always try to understand track preservation before you interpret the trackmaker's behavior. Put another way, consider the geological context of fossil footprints before jumping to biological conclusions. Failure to do so can lead to regrettable and embarrassing interpretations.

Here are a few examples of biological or behavioral interpretations of tracks that turned out to be highly dubious, or quite simply wrong, because those making the interpretations overlooked the geological context of the footprints and ignored available geological evidence. The first example is the trackway of a small bipedal dinosaur, named *Hopiichnus*, from early Jurassic deposits in Arizona. The trackway appears to be that of an animal with a remarkably long stride for the size of its prints, which would suggest that the trackmaker was running very fast. Our own reevaluation of this specimen indicates, however, that the trackway is very poorly preserved—a fact that was not mentioned in the original study. Thus the simplest explanation for the apparent long stride is that some tracks were eroded away, a geological rather than a biological explanation.

A second example of a faulty biological explanation of a phenomenon that was, in truth, geological pertains to the interpretation of sauropod trackways dominated by front footprints. An early explanation had the sauropods swimming, touching down more often with front than with hind legs (see chapter 5). But we have now demonstrated that the tracks are underprints, made on a layer below the original surface over which the animal walked. The front feet sank in deeper than did the hind feet in some cases. The hind foot underprints are thus faint or missing, having never been clearly impressed on the underlayer we see preserved today. Underprints are also tricky because they may be larger than true tracks, leading to a false impression of the trackmaker.

A third example of misinterpretation concerns parallel trackways. Such trackways are sometimes attributed to the passage of a herd or social group. However, parallel trackways may just as easily record different individuals passing along a preferred route, such as the shoreline of a lake. It has also been suggested that megatracksites are evidence of migration or of the congregation of large numbers of animals in a particular area during a relatively short span of time. The actual track record, however, provides no conclusive evidence of large-scale migration; we must recognize instead that the large concentration of tracks may also be the record of footprints that accumulated over a relatively long period of time, during a break in sedimentation.

Overall, many pitfalls challenge interpreters of fossil footprints. The best advice is thus for all trackers is to exercise caution and common sense.

Interpreting Track Assemblages

We have already explained that tracks of different types may occur together and that one can count the individual trackways at a particular site to arrive at a partial census of the ancient animal community of that place and time. In the field of ichnology, the relationship between the organism and its coinhabitors of a particular environment is important—so important that the mix of taxa that tend to be found together in trace fossil assemblages has been given a distinct name: ichnocoenose. A simple definition of an ichnocoenose is that it is an assemblage of traces produced by members of a particular animal community that characterizes a distinct environment. The term is used most widely in the study of invertebrate trace fossils, as many burrowing and crawling organisms actually live in the sediment and so are intimately related to particular environments that produce distinctive sedimentary rock layers.

Geologists studying many different trace fossil assemblages have noticed recurring patterns or associations of groups of traces (ichnocoenoses) with particular rock types. These rock types rich in trace fossils are called ichnofacies. An ichnofacies is a sedimentary deposit or facies in which particular trace fossil assemblages occur repeatedly. For example, certain types of burrow are found in sandstones representing high-energy shallow water, whereas other types are found in mudstones representing deeper, quieter water. Some ichnofacies are typical of marine sediment, others characterize freshwater deposits.

Although some tetrapod animals burrow underground, most trackmakers do not usually live in the sediment itself. Moreover, they tend to move around more than do most invertebrates, and so tetrapods may leave their tracks in many different areas, not just in one type of sediment. Despite these caveats, various footprint assemblages (vertebrate ichnocoenoses) often show recurrent patterns. This permits recognition of distinctive vertebrate ichnofacies. For example, certain types of tracks occur repeatedly in deposits representing arid sand-dune environments, whereas other types occur repeatedly in sediments that accumulated under more humid conditions. Some dinosaur tracks are prevalent in limestone deposits associated with low latitude, semi-arid climates with distinct seasons. Others are associated with swampy, wetland deposits that clearly represent humid environments (figure 1.14). Such differences indicate that in the past, as at the present time, the distribution of vertebrate animals was controlled to a significant degree by the environment.

Tracks Through Time

The history of vertebrate life spans three eras—Paleozoic, Mesozoic, and Cenozoic. Although fish originated very early in the Paleozoic (about 500 million years ago, near the end of the Cambrian period), tetrapods are virtually unknown until late in the Devonian period. A few reports of putative trackways of early Devonian and possibly even late Silurian age suggest a potential for trackways to shed light on the timing of the first emergence of tetrapods from water to land.

The traditional view is that vertebrate life began to proliferate on land early in the Carboniferous, during "the age of amphibians." At this time most living organisms were highly dependent on water, particularly for reproduction. Pictorial reconstructions of the Carboniferous usually show salamander-like animals inhabiting heavily vegetated, swampy environments. Unlike most of their descendants, these early amphibians were short-legged, long-bodied forms that

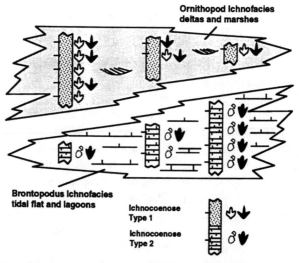

FIGURE 1.14 Trace fossil assemblages (ichnocoenoses) found in a particular sedimentary facies often exhibit a unique composition (recurrent pattern of species) and thus can be distinguished from those found in other sedimentary facies. This gives rise to the concept of distinct "ichnofacies" that broadly reflect different ancient environments.   *After Lockley et al. 1994.*

tended to sprawl and slither their way through life. Their characteristic wide trails and tail drag marks reveal this low-slung posture.

By mid-Carboniferous times, highly evolved amphibians had given rise to reptiles. One of the oldest known reptiles is *Hylonomus*, a lizard-like animal first found inside a fossilized hollow log in Nova Scotia. These creatures were less dependent on water and could invade drier, inland environments. This transition was accomplished by new adaptations, including improved (more erect) posture not seen in amphibians. The consequent changes in gait and locomotor ability show up in the track record as narrower trackways. For example, the Carboniferous track *Hylopus*, which is attributed to a *Hylonomus*-like animal, shows no evidence of tail dragging.

The Permian "age of reptiles" at the end of the Paleozoic was a prelude to the "age of dinosaurs," by which the Triassic, Jurassic, and Cretaceous periods of the Mesozoic era are loosely known. The transition from the "old" (paleo) to "middle" (meso) life era was, however, complex. The Permian marked the zenith of mammal-like reptiles, which are also sometimes referred to as reptile-like mammals. These distinctive tetrapods are generally thought to have been the first to develop thermoregulation (control of body temperatures). As thermoregulation physiologies improved, some tetrapods—including mammals, birds, and possibly some dinosaurs—eventually attained the controlled and sustained high metabolic levels that are loosely called "warm blooded."

Throughout most of the earliest of the Mesozoic periods, the Triassic, dinosaurs were unknown or inconspicuous, but their ancestors and close relatives among the archosaurs were on the rise, progressively displacing the mammal-like reptiles. The world was inhabited instead by a bewildering array of reptiles and an equally bewildering assortment of footprints. It was not until the Jurassic and Cretaceous periods that dinosaurs became dominant. Jurassic sediments yield familiar brontosaurs and their tracks, while Cretaceous sediments yield the bones of well-known duck-billed and horned dinosaurs and footprints to match. During both periods many other dinosaurs and other vertebrates (birds, mammals, amphibians, and non-dinosaurian reptiles) also existed and left a legacy of footprints.

To date, between 300 and 400 dinosaur track localities are known in the Colorado Plateau region. Many reveal evidence that has important implications for debates about dinosaur locomotion, behavior, and social or ecological interaction. At a significant number of these sites, interesting non-dinosaurian footprints are found alongside those of dinosaurs.

The Cenozoic "age of mammals" has also been referred to as the "age of birds." Both labels are appropriate in the region of the Colorado Plateau because the rocks already famous for producing the bones of numerous fossil mammals are also yielding a significant track record for birds. The era of "recent" animal life, the Cenozoic, consists of just two geologic periods: the Tertiary period (65–2 million years ago) and the very short Quaternary period, broadly equated with the Ice Age through the present. Even the Tertiary is, however, relatively short compared with Mesozoic and Paleozoic periods. Thus, an even-handed, equal-time treatment requires that the dinosaur story be given three times more space than that devoted to the Cenozoic record of birds and mammals. The Triassic, the Jurassic, the Cretaceous, and the Tertiary each receive a full chapter of discussion in this book.

Institutional Collections of Tracks

Only a decade ago tracks in the Colorado Plateau region were considered rare and somewhat insignificant. Now, however, it is known that tracks occur in remarkable abundance, sometimes providing the only significant record of vertebrate life in a particular geologic formation. The information and interpretations discussed in this book are thus, for the most part, very new indeed.

The majority of fossil footprints discussed here are held in the joint collection of the University of Colorado at Denver and the Museum of Western Colorado (abbreviated, CU-MWC). This collection has received the bulk of our own research attention over the last decade. The collection comprises about 600 individual slabs and replicas, including a particularly large group of late Triassic specimens.

To our knowledge, the CU-MWC collection is the largest in western North America. It is also very diverse—covering Paleozoic through recent specimens. Parts of the collection were used as the basis for a fossil footprint exhibit focusing on dinosaur tracks, which toured internationally. Specimens from this collection have been described quite extensively by us and others in the primary paleontological literature. In contrast, the largest collection of fossil footprints in the eastern United States—housed at Amherst College in Massachusetts—consists almost entirely of early Jurassic reptile tracks collected and studied in the early to middle nineteenth century by Edward Hitchcock, the founder of vertebrate ichnology. Although it is an historically important collection that has since been restudied by several scholars, it deals with little more than a single epoch of geological time.

Another interesting and extensive collection in the western United States is housed at the Raymond Alf Museum at the Webb School (a private high school) in Claremont, California. This collection includes a large number of Permian sandstone slabs from the Grand Canyon area. Although the Alf collection also contains other tracks of Mesozoic and Cenozoic age, the specimens are not numerous and they are not yet fully studied or documented.

The recently excavated Permian tracks from the Abo Formation of southern New Mexico, now at the New Mexico Museum of Natural History and Science, constitutes another world-class collection that represents but a single epoch and that has yet to be studied in detail. Similarly, the collection of dinosaur tracks at the Tyrrell Museum of Palaeontology in Drumheller, Alberta, focuses mainly on Cretaceous footprints that were excavated in western Canada. Other noteworthy collections are housed at the Museum of Northern Arizona in Flagstaff, the College of Eastern Utah Prehistoric Museum in Price, Brigham Young University in Provo, the University of Utah in Salt Lake City, and the University of California at Berkeley. A number of large museums in the eastern United States have also accumulated noteworthy collections of western tracks. Among the most frequently cited are specimens from the collections in the Smithsonian in Washington, D.C. and the American Museum of Natural History in New York City.

The appendix to this book lists the major fossil footprint sites and institutional collections in the western United States that are open to the public.

Conservation and Preservation

Like dinosaur bones and other body fossils, footprints are a valuable and nonrenewable natural resource. In recent years we have noticed a significant increase in the number of track-bearing slabs appearing in the commercial fossil market. Although many for-profit collectors are responsible in their activities and work cooperatively and legally with the scientific community, the movement of fossil tracks into private holdings can pose significant problems for further research. Most land management officials are not paleontologists, nor are they physically able to police the large tracts of land where fossil footprints are found. And even tracks that are recorded may need to be studied again at a later date, when new questions may be posed and new technologies become available. This means that in many cases discoveries still in situ need to be fully documented, replicated, or, if in danger of imminent natural or human-induced destruction, collected to preserve their current scientific value and potential.

There are three main approaches to preserving a permanent record of tracks or any other fossil. First, one can provide scientific documentation in the form of written descriptions, maps, photographs, and other illustrations, as done in this book. The second strategy is to make replicas of the specimens; the third is to actually collect the specimens and reposit them in a museum. The latter two approaches have also regularly been employed in our own studies—but not in all cases.

The decision about whether or not to collect a specimen depends on several factors, including size, quality of preservation, rarity, and scientific value. Unlike bones, which have a finite size and which are almost always collected to prevent them from deteriorating, tracks form part of continuous trackways that cannot be even partially excavated (removed rather than uncovered) without causing significant damage to the site. Moreover, collecting several consecutive large tracks, and the intervening rock surface, is a massive undertaking that would strain the resources of almost any institution. For this reason it is much easier to make replicas of the track surface. This is usually done by making a rubber mold of the desired track, trackway segment, or portion of tracksite surface and then making a plaster of paris or fiberglass replica from the mold.

Replicas and molds provide paleontologists with a permanent record of three-dimensional specimens. In most cases such replicas will outlast the original material in the field, which is subject to deterioration or loss through weathering, erosion, burial, theft, or vandalism. At many sites pieces of track-bearing strata become detached from the main outcrop during the course of normal weathering and erosion. Once removed from their original position, information on their orientation and even the layer that they came from begins to be lost—though in some cases we can put the pieces back in place, and so record information in proper context. Usually if such detached slabs are portable, and the investigators hold the requisite permits, they are removed to safekeeping in a designated museum repository, where they are protected from further deterioration. The CU-MWC collection contains such a mixture of track replicas and original specimens.

Some paleontologists express extreme consternation about commercial collecting. Others view commercial collectors as real or potential allies who help find and develop valuable sites. In our experience, working at over three hundred sites, only a few have been vandalized or plundered; tracks have been illegally or thoughtlessly removed from less than 5 percent of these sites. We encourage anyone in a position to assist with documentation, replication, conservation, or acquisition of specimens by and for museums to do so. Such activities further the cause of conservation and will lessen concern about future acts of vandalism.

Ultimately, it is not the duty of universities and museums to protect the physical resource of tracks or fossils in the field, even though such institutions can and should help considerably. The responsibility, rather, is borne by various land management agencies and landowners (state, federal, and private). However, the scientific community can help in the conservation and preservation process by providing a permanent and comprehensive written record—and by obtaining specimens and replicas for various repositories. We have attempted to do this with respect to the fossil footprint collections of the Colorado Plateau region—in some cases even prevailing upon private collectors to donate specimens or at least make them available for replication. We regard this as an important first step in the conservation and preservation process.

Attempts should also be made to protect important field sites and to educate the public about their significance. To this end, we have withheld the exact location of most sites discussed here. Such information has been (and will continue to be) shared with research professionals, teachers, and land management officials, on a request basis.

Tracker looking at markings that resemble horse tracks, Paleozoic strata of the Grand Canyon region. Artwork by Paul Koroshetz.

2

ANCIENT TRACKS: THE PALEOZOIC ERA

One generation passeth away, and another generation cometh: but the earth abideth forever.

ECCLESIASTES 1: 4

■ Sometime in the middle of the Paleozoic era, probably during the Devonian, the first tetrapods began leaving trackways on land. These pioneers were amphibians that had only recently evolved from lobe-finned fish. Although their fossil record is sparse, it nevertheless represents a crucial stage in vertebrate evolution. The transition from fins to legs and from water to land may have been a small step for an amphibian, but it was a giant step for the tetrapods, which went on to dominate the continents for the next 400 million years.

Devonian tracks are very rare and somewhat controversial. Examples have been reported from Scotland, eastern North America, Brazil, and Australia, but not everyone accepts that these poorly preserved specimens are true tracks. They could be deceptive track-like features. No Devonian footprints have yet been reported from western North America. This absence is not surprising, given their extreme rarity worldwide, but it also may reflect a lack of suitable shoreline deposits at that time.

In the Carboniferous (360–286 million years ago) footprints appear to be quite abundant, and this is when the track record begins in most regions of the world. Reptiles had already originated by this time, though even specialists have difficulty in distinguishing them from their amphibian contemporaries. Thus, many Carboniferous tracks could be attributed to either amphibians or early reptiles.

Western Traces in the "Age of Amphibians"

Traditionally, the Carboniferous (meaning, coal bearing) is referred to as the "age of amphibians" or the "age of coal swamps." Carboniferous coal is found in abundance in eastern North America, where it fueled the industrial revolution. Carboniferous tracks have been found as a by-product of coal mining in many eastern states, including Alabama, Georgia, West Virginia, Ohio, and Pennsylvania. Often such tracks are seen in the ceilings of coal mines where the underlying layers have been removed.

In the western states, however, coal accumulations are virtually unknown in strata of this age. A different habitat yielded a different and evidently sparser track record in the West. We know of a few Carboniferous sites in Kansas, Oklahoma, Colorado, and Arizona but must accept that at present they are poorly documented. In future they may turn out to be abundant.

The track record overall tends to be poorer the further back one searches in time. This may in part be a function of a relative scarcity of track evidence from Carboniferous deposits. The poorer record may also reflect preservation problems that impede the recognition and identification of predominantly small tracks. However, lack of knowledge is surely also a function of lack of detailed study. As recent discoveries of Permian and Mesozoic tracks have shown, there is generally no shortage of material for the diligent tracker to find and study. We therefore predict that the data base will greatly expand as new Carboniferous sites are found and studied.

Probably the best-known Carboniferous footprints from the western states are those described by Charles Gilmore. These footprints occur in the Wescogame Formation, a part of the Supai Group of the Grand Canyon area, and they were initially discovered by Charles Schuchert in 1915. Gilmore was sent to the Grand Canyon in the 1920s to collect specimens of these tracks and other abundant Permian footprints for the Smithsonian Institution and to develop an outdoor interpretive exhibit of Paleozoic tracks alongside Hermit Trail. His mission was successful, and in a series of papers he reported on a large number of sites and trackways. By doing so, he added considerable knowledge of the Paleozoic track record, ranging from the Coconino Sandstone of the middle Permian to the Hermit Shale of the early Permian and then back to the Supai Formation of the late Carboniferous (Pennsylvanian).

The Supai Formation has yielded tracks (figure 2.1) attributable to a number of large trackmakers, for example, *Anomalopus* and *Tridentichnus*, which appear to have had four and five toes, respectively, on their hind feet. The Supai beds have also yielded smaller, slender-toed and lizard-sized tracks referred to as *Stenichnus*. All these tracks require further study—having received little scientific attention since Gilmore's day.

In 1932 a trackway was reported by Edward Branson and Maurice Mehl from the Carboniferous Tensleep Sandstone of Wyoming. They named the trackmaker *Steganoposaurus belli*, which loosely translates as "beautiful amphibian track." These authors interpreted the trackmaker as being an animal "somewhat less that three feet in length" with webbed feet. *Steganoposaurus* tracks are similar in

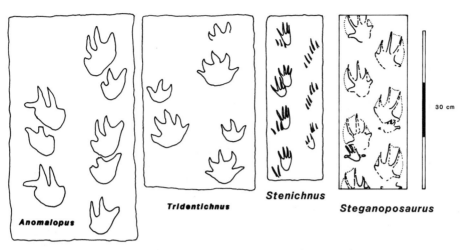

30 cm

Anomalopus

Tridentichnus

Stenichnus

Steganoposaurus

FIGURE 2.1 Carboniferous tetrapod tracks from Arizona and Wyoming, including *Anomalopus*, *Tridentichnus* and *Stenichnus* from the Supai Group of Arizona *after Gilmore 1927* and *Steganoposaurus* from Wyoming. *After Branson and Mehl 1932.*

many respects to *Tridentichnus* from the Supai Formation, but little is known of the trackmaker's identity. In recent years Carboniferous trackers suggested that these "beautiful amphibian" tracks were made not by amphibians but by primitive *Hylonomus*-like reptiles. This change in interpretation is justified, at least in part, by the fact that *Steganoposaurus* tracks have slender, pointed digit impressions, whereas amphibians normally have blunt, rounded toes.

Carboniferous Tracks: Amphibian, Reptile, or Other?

Could *Steganoposaurus* indeed have been made by a reptile? Consider: much of the Tensleep Sandstone consists of wind-blown sand dunes that were probably not a favorable habitat for amphibians. Thus, Branson and Mehl's original interpretation of an amphibian trackmaker should be viewed with caution. Moreover, it is also probable that their original web interpretation is wrong.

Despite the "age of amphibians" label for the Carboniferous and the fact that some amphibians alive today do have webbed feet, the Carboniferous track record reveals very few examples of footprints with web impressions. This is in part because webs are soft and delicate and often do not exert enough pressure on the ground to leave impressions. Many reports of post-Paleozoic footprints with web impressions have later been shown to be outright incorrect, dubious, or better explained as features that result from imperfect preservation. The lack of many convincing examples of web traces is perhaps because webs are associated with aquatic animals rather than those that more frequently walk on land and because a web-footed animal may not always splay its toes to extend the web when walking (see chapter 3). Indeed, if the post-Paleozoic track record is reliable, amphibians simply do not leave many tracks, especially in sand dunes.

Despite all these difficulties in distinguishing amphibians from reptiles, many Paleozoic tracks in museums are boldly labeled "amphibian footprints."

Paleozoic Horses?

During his several years of study on Paleozoic footprints from the Grand Canyon, Charles Gilmore also reported some strange "pseudo-track-like markings" from the Supai Formation. These bear a remarkable resemblance to tracks made by horses' hooves. (See the frontispiece figure to this chapter.) Gilmore reported that these markings had been known to the local Supai Indians for some time, who interpreted them as tracks made by a band of horses. Gilmore noted that these were not the first horseshoe-shaped markings that had been interpreted as tracks. The famous ichnologist Edward Hitchcock had actually named a similar set of markings from Jurassic beds in Connecticut in 1858.

In Gilmore's time it was already known that there were no horses around in the Paleozoic. Gilmore concluded that the markings were not attributable to vertebrates and could instead be explained as the natural staining and differential weathering of the rock. Despite Gilmore's conclusion, the eminent geologist Edwin McKee reported over a half century later, in 1982, that he had observed similar but not identical horseshoe-shaped traces in the Esplanade Sandstone of the Carboniferous. Because vertebrates are known to have been present at that time, McKee interpreted these traces as "trackways of a quadrupedal animal."

Similar horseshoe-shaped markings have been reported in a number of younger Mesozoic deposits from North America and other parts of the world. In some cases they have been interpreted as vertebrate tracks, thus providing fuel to creationists intent on discrediting fossil evidence. But careful studies show that most are invertebrate traces or features produced by currents. A review of all known reports of this type suggests that a surprising number of invertebrate traces have been misinterpreted as vertebrate tracks. We conclude that, in most cases, the vertebrate track interpretations are dubious at best and that each example should be examined carefully and judged on its own merits. In the case of the Supai Formation, of course, we agree with Gilmore that the markings are not vertebrate tracks—and they were definitely not made by horses!

This reflects the outdated, but still prevalent, opinion that any Paleozoic tetrapod was primitive, and therefore at the lowest grade—that is, an amphibian.

Carboniferous tracks are also known from at least three localities in Colorado, including two in the Pennsylvanian-aged Minturn Formation. The first discovery was made near Dotsero in 1953 and the second on the flanks of Hermit Peak in the Sangre de Cristo Mountains in 1986. As shown in figure 2.2, the first discovery was of a rather incomplete trackway, with no obvious traits to distinguish its maker as amphibian or reptile. The second, however, revealed a single trackway with an obvious tail drag mark. This feature indicates a short-legged, sprawling animal built somewhat like a salamander. Because both sets of tracks were found in sediments that represent coastal wetlands and delta environments (not dry, sand dune settings) it is possible that one or both were made by amphibians.

The Minturn Formation footprints are relatively easy to date accurately. The track-bearing layers of terrestrial sediment alternate with near-shore marine layers containing well-known, easily dated fossils, about 300 million years old. This same formation has also yielded the tracks of horseshoe crabs, also known as limulids and related to the extant genus *Limulus* that typically congregates along beaches or in estuaries (figure 2.3). These limulid tracks are of particular interest because they typically occur right along shorelines, often in association with vertebrate tracks, and as a result can be confused with vertebrate footprints.

A classic example of this confusion occurred in the 1930s, when the paleontologist Bradford Willard described some Devonian tracks from Pennsylvania and called them *Paramphibius* (meaning, toward an amphibian). He posited their

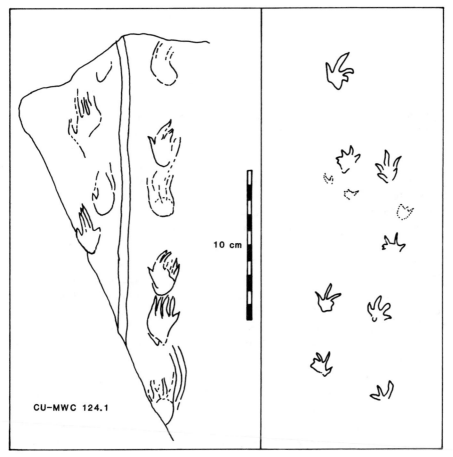

FIGURE 2.2 Carboniferous tetrapod tracks from Colorado. *Left:* trackway with tail drag from the Minturn Formation (after Lockley 1989). *Right:* tracks from the Belden Formation (after Houck and Lockley 1987). CU-MWC 124.1 and USGS D724 respectively.

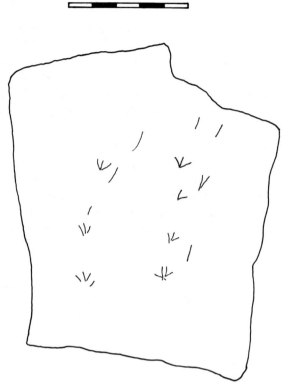

5 cm

FIGURE 2.3 Horseshoe crab tracks from the Carboniferous of Colorado (Minturn Formation, near Vail).

maker as being an animal intermediate between fish and amphibians and coined the name Ichthyopoda (meaning, fish with feet). In Willard's view, the *Paramphibius* tracks were important because they were traces of the missing link! It turns out that Willard was wrong. In a classic and exemplary study published in 1938, Kenneth Caster effectively proved that *Paramphibius* tracks were made by horseshoe crabs by finding their actual remains in the same formations as the tracks. The name *Paramphibius*, therefore, was ill-conceived and clearly based on Willard's misconceptions. Such traces have been given a number of names over the years—reflecting their abundance and diverse shapes. As we shall later detail, the tracks of horseshoe crabs are found in many western deposits from Carboniferous to Jurassic, and they often occur on the same surfaces as vertebrate footprints.

There is at least one other report of Carboniferous tracks from Colorado, namely *Baropezia (Brachydactylopus) fontis* from the Fountain Formation of the Front Range area (figure 2.4). This specimen indicates a larger animal than those from the Minturn Formation, which existed at about the same time. This distinctive track was given its cumbersome, but descriptive, Latin name to indicate that it represented a "heavy-footed (broad-toed) creature from the Fountain Formation." The ichnogenus name in parentheses indicates that the original name, *Brachydactylopus*, is considered a synonym of a preexisting name, *Baropezia*. We take this synonymy one step further and note that this type of track is similar to the track referred to as *Limnopus* (meaning, lake foot or footprint from a lake environment), which we will discuss later in this chapter. Again, like the other tracks from the Carboniferous of Colorado, *Baropezia* was found in sediments representing wetter alluvial environments—not in desert sand dune deposits. Such tracks are generally attributed to amphibians.

Tracks of the Arid Permian

Unlike the Carboniferous, which is traditionally viewed as a humid time, the following Permian period brought an arid climate to much of North America and Europe. The aridity was, at least in part, a result of the harsh continental climates on the supercontinent known as Pangaea II. As discussed in the next chapter, the world's continents coalesced into one giant land mass during the late Paleozoic and early Mesozoic, thus giving rise to unique geographic and climatic conditions. Throughout much of western North America there is good geological evidence for arid Permian environments. In Texas, for example, abundant small reptile tracks have been recovered from middle Permian sediments representing lowland playa lake environments that signal alternating wet and dry seasons.

As reptiles evolved the ability to lay shelled eggs, they were able to breed and survive in drier environments away from the swamps and waterways so vital to amphibian life. Thus, although the extent of wet environments decreased, the reptiles followed plants and insects, colonizing virtually the entire landscape and spearheading the colonization of drier land in the continental interiors.

Probably the most obvious evidence of arid conditions in the Permian comes in the form of fossil sand dune deposits, attesting to Sahara-like desert environ-

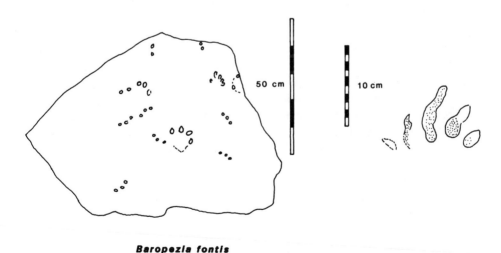

Baropezia fontis

FIGURE 2.4 *Baropezia fontis* track slab from the Fountain Formation, Colorado, with detail of a single track of this probable amphibian from a companion slab (specimens in University of Colorado at Boulder collections).

ments known as "ergs" or sand seas, or to extensive coastal dune fields in some cases. Such dune deposits—often referred to as "eolian" (meaning, of the wind)—include much of the Esplanade Sandstone and the Coconino Sandstone of the Grand Canyon area, the De Chelly Sandstone of the Monument Valley area, the Cedar Mesa Sandstone of eastern Utah, and the Lyons Sandstone of Colorado. These are mostly dated to the early or middle Permian.

"Stone Tracks" of the Coconino Sandstone

Perhaps surprisingly, all these Permian fossil dune deposits have yielded abundant footprints, including tracks attributed to mammal-like and lizard-like reptiles as well as smaller traces attributed to spiders and other arthropods. Perhaps even more puzzling than the abundance of tracks in desert settings is how they were preserved in what appears to have been an environment of dry, shifting sands. Various authors have suggested that the clear, crisp tracks could only have been made after the sands had been wetted by dew. Others postulate that fine layers of wind-blown dust draped the dunes, impregnating the spaces between sand grains and thus providing a more cohesive surface on which tracks could be made.

Among the most famous of the Permian fossil footprints are those from the Coconino Sandstone of the Grand Canyon area, which can now be found in countless museum displays and collections. Following a Smithsonian expedition led by Charles Gilmore in the 1920s, numerous track-bearing slabs have been extracted by museum scientists and commercial collectors. One of the best collections is on display at the Raymond Alf Museum at the Webb School in Claremont, California. Unfortunately, many other spectacular specimens are freely traded on the commercial market.

When Richard Swan Lull first described these trackways, he named them *Laoporus* ("stone tracks"), as he was unable to attribute them to known trackmakers. However, many of his contemporaries attributed *Laoporus* tracks to amphibians—an identification frequently repeated in museum exhibits. The notion of an arid desert crawling with amphibians is contradictory, to say the least; recent work has suggested that the tracks are probably mainly those of a group of mammal-like reptiles known as caseids.

Despite this new interpretation, a few workers contend that some Coconino footprints were made underwater by an amphibious vertebrate, not on the dry slopes of dunes. This interpretation is based on unusual trackways that abruptly begin and end or that show sudden changes from forward progression to sideways movement (figure 2.5). Experiments with living animals suggest that such unusual trackways are sometimes produced by salamanders maneuvering in shallow water; however, such experiments do not provide conclusive, or even convincing, evidence that amphibians made the Permian trackways. The studies in question also failed to comment adequately on the geological evidence for dunes and the previous identification of the trackmakers as mammal-like reptiles.

Trackways found in fossil sand dune deposits have generated much interest over the years. One of the most common observations is that the tracks often have bulges or sand crescents on one side, thereby proving that they were made

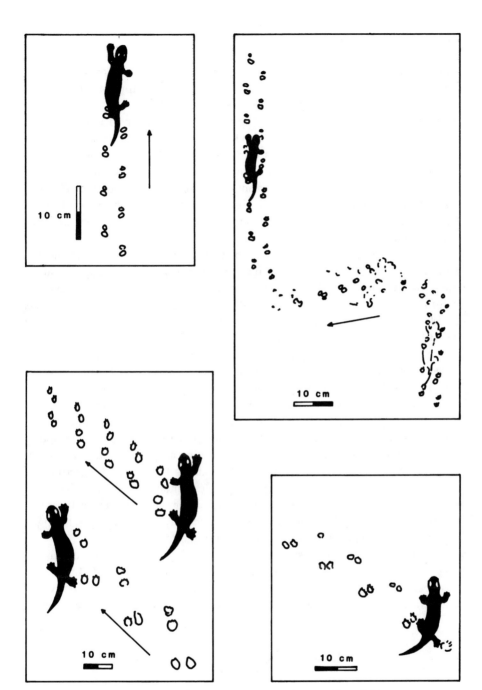

FIGURE 2.5 Unusual tracks from Coconino sandstone, Grand Canyon area, show animals progressing at various angles to slopes of fossil dunes. The arrows represent direction of travel. (Compare with figures 2.6 and 2.13.) *Based on specimens in the Raymond Alf Museum; after Brand and Tang 1991 (which interpreted the patterns as swimming marks).* Brand (1997), in a creationist textbook, has suggested a link between these unusual trackways and the behavior of animals during a catastrophic biblical-style flood. We cannot support this interpretation of the geology and ichnology of the Coconino track-bearing sediments.

on inclined surfaces. Typically these sand crescents—also sometimes referred to as impact rims—are situated behind the rear or "heel" of footprints, showing that the animals were progressing upslope. It seems that this type of upslope track-way is the most common and usually the best preserved. Even so, trackways that indicate downslope progression are also known, as are a number that show side-ways or oblique movement across dune faces. These sideways or transverse trackways are especially interesting because the tracks often point upslope while the trackway crosses the slope horizontally or obliquely (figure 2.5). There is, however, a modern analog. We have observed fresh trackways of lizards, with tail drag marks, that run transversely across dune faces (figure 2.6), leaving indi-vidual tracks that point uphill.

The unusual Coconino trackways also often reveal very regular footprint con-figurations reminiscent of loping, trotting, or galloping gaits. This suggests that the trackmakers sometimes adopted rapid or running gaits when attempting to nego-tiate dunes. Indeed, Don Baird suggested that one distinctive trackway type, *Dolichopodichnus* (see chapter 4), which Charles Gilmore originally attributed to a lizard-like animal, was nothing more than a *Laoporus* trackway made while the trackmaker was running. Using R. McNeill Alexander's formula for estimating speed from trackways, it is possible to calculate that a typical *Laoporus* trackway (for example, figure 2.8, left) was made by an animal moving at about 1 kilometer per hour. The *Dolichopodus* trackmaker was moving at twice that speed. Such examples are probably the oldest known trackways that could be soundly inter-preted as evidence of running or least of a faster pace than normal walking speed.

For all these reasons, we contest the swimming amphibian interpretation of *Laoporus* trackways. Even so, we strongly endorse the value of experimental work with modern trackmakers. Experimenters should remember that even though a modern animal makes a trackway similar to that seen in the fossil record, alternative explanations for the fossil trackway may actually be stronger, especially if the depositional environment so suggests.

A final argument against the interpretation of a subaqueous origin for *Laoporus* tracks is that they are associated with many invertebrate traces that have been attributed to spiders and scorpions. Two of the better-known inverte-brate traces found in the Coconino Sandstone are *Octopodichnus* ("eight-footed

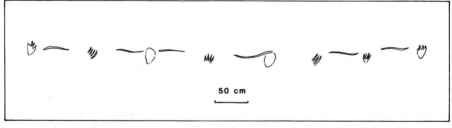

FIGURE 2.6 This modern lizard trackway shows horizontal travel along the contours of a dune face.
Drawn from photograph taken by the authors near Dinosaur National Monument, Utah.

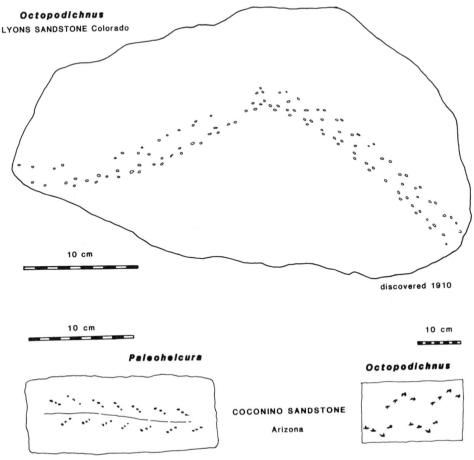

Octopodichnus
LYONS SANDSTONE Colorado

10 cm

discovered 1910

10 cm

Paleohelcura

COCONINO SANDSTONE

Arizona

10 cm

Octopodichnus

FIGURE 2.7 Invertebrate traces from the Coconino Sandstone, Arizona, and the Lyons Sandstone, Colorado.

trace") and *Paleohelcura* ("ancient wound or mark"). Most ichnologists agree that both traces were made by arthropods of some type, but interpretations range from spiders to scorpions to crustaceans (figure 2.7).

In a series of papers published from 1939 to 1961, Lionel F. Brady undertook a series of experiments with small living arthropods. He determined that *Paleohelcura* tracks are very similar to those made by living scorpions and that *Octopodichnus* might also be attributed to a scorpion-like animal. Brady also found that the same scorpions made different traces at different temperatures, depending on whether they were torpid or warmed up. Raymond Alf, founder of the museum in Claremont that now bears his name, continued this tradition of experimentation with living arthropods. In 1968 he showed that certain traces from the Coconino Sandstone were very similar to those produced by large tarantula spiders. It is interesting to note that Alf overlooked or at least failed to cite Brady's work and therefore did not comment on the similarity between the spider tracks and *Octopodichnus*.

Recently, Christa Sadler has found these same arthropod traces in eolian deposits beyond the Coconino Sandstone. She has described *Octopodichnus* and

Paleohelcura from the De Chelly Sandstone. This formation is another Permian deposit, famous for forming the towering features of Arizona's Monument Valley. At one location in Canyon de Chelly, a tower of De Chelly Sandstone is named Spider Rock. This is not a reference to *Octopodichnus*, but an interesting coincidence nonetheless. As this book was going to press, we are undertaking a study of three new vertebrate tracksites in the De Chelly Sandstone that contain *Laoporus*-like tracks and abundant *Octopodichnus* at dozens of different levels.

The weight of biological and paleoenvironmental evidence pertaining to the Coconino Sandstone and De Chelly Sandstone now strongly argues against a subaqueous origin for the tracks. There is very little evidence to suggest extensive bodies of water in Coconino deposits. Even those who support the swimming interpretation for *Laoporus* tracks cannot show conclusively that these vertebrate trackmakers were amphibians or that dry-land explanations must be abandoned.

The Lyons Sandstone: Twin of the Coconino

Laoporus tracks are also the dominant track type in the Lyons Sandstone of Colorado. This too is a Permian sandstone of desert origin that has yielded many well-preserved trackways, including some with unusual configurations. The Lyons Sandstone has also produced the invertebrate traces *Octopodichnus* and *Paleohelcura*. *Octopodichnus* was discovered in the Lyons Sandstone in 1910, but an actual specimen is illustrated here for the first time (figure 2.7).

The strong similarities in the Coconino, De Chelly, and Lyons traces are no coincidence. The Lyons Sandstone is about the same age as the Coconino Sandstone, and both represent dune-field environments. It is not surprising that both areas were inhabited by the same types of animals. The rock types and traces are so similar that a distinct *Laoporus* ichnofacies has recently been designated. Similar track assemblages (ichnocoenoses) are reported from desert sandstones of Permian age in Europe. Based on extensive experience in studying Permian traces, our colleague Heinz Kozur informs us that the concept of a *Paleohelcura-Octopodichnus* ichnofacies would also be an appropriate way to characterize the traces found in these desert dune deposits. He agrees that *Laoporus* tracks are intimately associated with sand dune deposits in the western United States, but he notes that in European rocks of the same age they also occur in other types of arid deposits representing playa lakes that adjoined the dune fields.

The habitat and ecological community that marks the *Laoporus* ichnofacies may not only have been widespread but long-lasting. It appears to be present in a very similar form in much younger sand dune and playa lake deposits of Jurassic age (see chapter 4).

For many years trackers have known that *Laoporus* is very similar to Permian tracks from Europe that go by the name of *Chelichnus* (meaning "tortoise tracks"). The name *Chelichnus* was coined in 1850, long before Lull introduced the name *Laoporus*. Our colleague Dr. Hartmut Haubold informs us that *Chelichnus* and *Laoporus* are essentially the same track type. This implies that *Laoporus* could soon be suppressed and replaced by the name *Chelichnus*, which has historical priority. We agree with Haubold's assessment of the similarity between the tracks from Europe and North America and also agree that the rules of nomenclature allow such a substitution of names. It is unfortunate, however, that the older name suggests the trackmaker was a tortoise when everyone now agrees that the trackmakers were mammal-like reptiles.

It should also be noted that if future work established *Chelichnus* as the valid name for *Laoporus* then trackers may wish to substitute the term *"Chelichnus* ichnofacies" for the recently introduced term*"Laoporus* ichnofacies" [one could even talk of a *"Chelichnus (Laoporus)* ichnofacies" to underscore that the names are synonymous]. Whatever terminology is eventually used, the message is the same. This type of ichnofacies, dominated by the tracks of mammal-like reptiles, is widespread in the Permian of both Europe and North America. Our book is about the western United States so the name *Laoporus* had to be applied, at least initially. The future adjustments in terminology anticipated by Dr. Haubold and ourselves will only serve to underscore the widespread distribution and abundance of cosmopolitan trackmakers in arid Pangaean environments.

The Lyons Sandstone has been extensively quarried for building stone, mainly from the quarries near the town of Lyons. Many buildings on the Boulder campus of the University of Colorado are built of this stone. Many track-bearing slabs have fallen into the hands of private collectors, but some have been preserved in museum collections. As so few have ever been described or even illustrated, we have chosen to illustrate several specimens here (figure 2.8 and 2.10).

The type specimen of *Laoporus coloradensis* (figure 2.8) shows regular walking progression with well-defined front and back footpints—as do many of the examples from the Coconino Sandstone (figure 2.9). We have studied several other trackway slabs with interesting configurations. One slab shows three trackways that are more or less parallel and only a short distance apart (figure 2.10). The trackways lack well-developed sand crescents that would otherwise suggest progression up steep slopes. Subtle rims do, however, indicate that the animals may have been moving up a gentle slope. Examples in figures 2.8 and 2.9 indicate straightforward progression.

How might one interpret figure 2.10? This pattern could indicate a group traveling together at the same time. If this were true, then this specimen would be the earliest example of gregarious behavior demonstrated from trackway evidence. We know of at least one additional specimen, from the Coconino Sandstone, that also shows a large number of parallel trackways suggestive of a gregarious group (figures 2.11 and 2.12). On the other hand, we note that each trackway has a different type of preservation, ranging from toe-only impressions to full footprints or indistinct oval indentations. This may suggest that the trackmakers passed by at different times, when substrate conditions were different. If this were the case, then the gregarious behavior scenario would no longer hold—unless one countered that, even in a small area, track preservation can vary as a result of subtle differences in the substrate. In either case, this particular slab is a reminder that one must be cautious in drawing behavioral conclusions from tracks.

The final Lyons example we have illustrated, figure 2.13, is another University of Colorado specimen that shows one trackmaker progressing directly upslope and another cutting obliquely across. In both cases, prominent sand crescents are clearly visible behind the tracks, suggesting that the slope was quite steep. The obliquely oriented trackway is similar to several that we referred to as "unusual" in the Coconino Sandstone. This leads to an obvious question: The Lyons trackmakers behaved in a similar fashion to those in the Coconino, so were they swimming or were they moving in a dune-climbing gait?

To be consistent in our interpretations we must opt for the climbing gait interpretation. If anyone were to suggest a swimming interpretation, they would have to also propose that the Lyons Sandstone was deposited, at least in part, underwater. This is not consistent with studies of the sandstone in the areas where the tracks were found. In these areas, geologists have identified sand

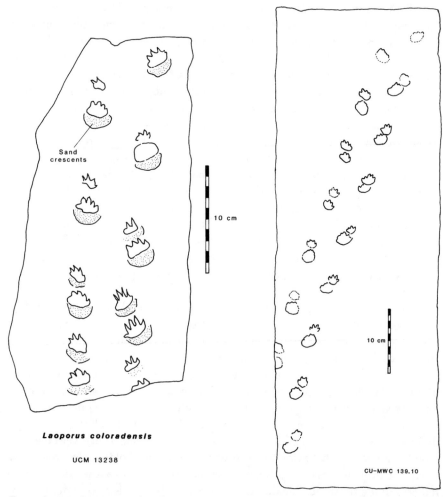

FIGURE 2.8 *Laoporus coloradensis* is the track of a kind of mammal-like reptile. *Left:* type specimen. *Right:* a similar trackway. Both are from the Lyons Sandstone, Colorado.

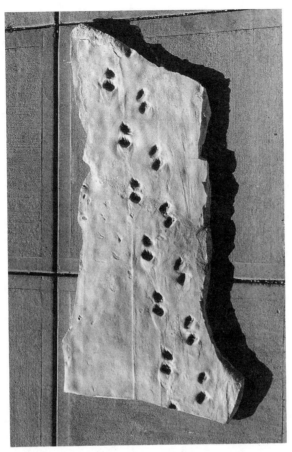

FIGURE 2.9 A typical *Laoporus* trackway from the Coconino Sandstone shows normal walking progression of a mammal-like reptile. This specimen is a replica of a slab in the Potomac Museum Group collection.

avalanche features and rain drop impressions, both of which show that the ancient sand dunes were exposed to the atmosphere and not submerged beneath bodies of water.

Interpreting Tracks and Track Habitats

We still know surprisingly little about the Coconino and Lyons tracks and traces. Despite the fact that large collections of Coconino Sandstone tracks exist in the Smithsonian Institution, the University of California Museum of Paleontology, the Raymond Alf Museum, the Museum of Northern Arizona, and the Potomac Museum Group collections, very little has been written since the contributions of Gilmore between 1926 and 1928.

When viewed as a whole, the Coconino and Lyons tracks and traces are interesting because they indicate an animal community in which the creature responsible for making *Laoporus* tracks (a mammal-like reptile) was very dominant. From an ecological perspective, such marked dominance and low diversity is a

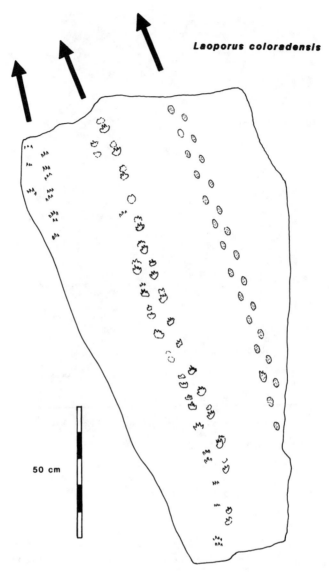

Laoporus coloradensis

50 cm

FIGURE 2.10 Three parallel *Laoporus* trackways from the Lyons Sandstone (CU-MWC 139.11) could indicate gregarious behavior. Note the different styles of preservation, from toe impressions only to full tracks and oval depressions.

sign of an opportunistic species in a stressed environment. This interpretation appears to be consistent with the inference that the environment was a desert. The scorpion-like and spider-like traces also support this interpretation. But what did the vertebrates that made the *Laoporus* tracks eat?

One possible answer may be found in an interesting slab recovered from the Coconino Sandstone in Arizona, by Jon Kramer. This specimen (figures 2.14 and 2.15), now in the Minnesota Science Museum, shows the trackway of a large invertebrate that ends abruptly at the point where a vertebrate trackway crosses the slab. It is tempting to interpret the termination of this trail as evidence that the vertebrate ate the hapless invertebrate. If this interpretation is correct, then this specimen could be cited as evidence that desert-dwelling vertebrates actually fed on some of the invertebrates responsible for making traces in the Coconino Sandstone.

There is another interpretation, however, and it is one that may disappoint the dozen or more paleontologists who have seen the specimen and gone along with the initial vertebrate-eats-invertebrate interpretation. According to our colleague Heinz Kozur, some European trackers consider that such trace fossils sometimes end abruptly because the trackmaker was some kind of insect that simply flew off. Close examination of the trackway indicates that the invertebrate trackmaker was indeed prone to making abrupt changes in direction,

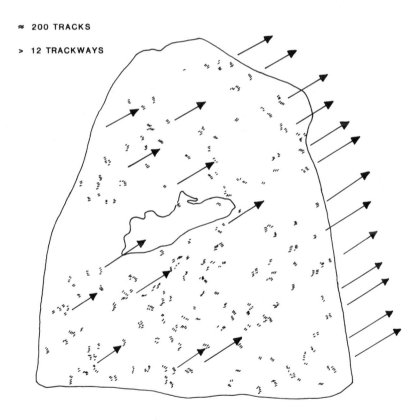

PARALLEL TRACKWAYS   COCONINO SANDSTONE

Laoporus   SELIGMAN   ARIZONA

≈ 200 TRACKS

> 12 TRACKWAYS

Potomac Museum Group Specimen

FIGURE 2.11 This map of a slab of Coconino Sandstone shows at least a dozen parallel trackways of *Laoporus*. Slab width is about one meter.   *Courtesy of Jon Kramer and the Potomac Museum Group.*

FIGURE 2.12 A slab of Coconino Sandstone with parallel *Laoporus* trackways suggests gregarious behavior among mammal-like reptiles. (Compare with Figure 2.11.)

including a sharp turn to the left before it "took off." Regardless of whether the trackmaker was eaten or simply flew away, this is the first discovery of this type of trace fossil in the Coconino Sandstone, so the find adds to the list of arthropod trackmakers that inhabited Permian sandstone environments in the western United States.

A review of the animals that inhabit modern desert environments reveals a number of insects (burrowing wasps, flies, beetles, crickets, ants, and termites), along with spiders and scorpions and occasional millipedes and crustaceans. The most common modern vertebrates are lizards and small mammals (rodents). The overall picture of modern desert communities is therefore quite analogous to that indicated by the Permian tracks and traces.

Not all Permian tracks in the western states come from ancient dune deposits. In some areas they are found in what we call alluvial fan sediments. Alluvial sediments occur at the present time mostly along the flanks of mountain ranges that are bounded by active faults. Such sediment fans are usually a sign of rapid erosion in semi-arid settings where there is little vegetation to stabilize soil and sediment and where there is a significant level of tectonic activity. The sediments consist of thick wedges of coarse sand and gravel that are washed rapidly out of the mountains by periodic rains and flash floods. Further out in the basin, at the toe of such fans, floodwaters and finer muddy and silty sediments accumulate in ephemeral lakes or small pans.

Some of the best examples of modern alluvial fans are found in Nevada in what is known as the Basin and Range province. An especially good ancient example of a system of Permian alluvial fans is known as the Cutler Group, and it accumulated on the western flanks of the ancient Rocky Mountain uplifts in the area that is now western Colorado and eastern Utah. In 1957 tracks were reported from these sediments and given the name *Limnopus cutlerensis*. (*Limnopus* means "lake foot.") It is thought by some to represent the tracks of a temnospondyl amphibian (figure 2.16).

The name *Limnopus* was coined in 1894 by the famous paleontologist Othniel C. Marsh. Marsh applied it to tracks he discovered from Carboniferous coal beds of Kansas. Other examples of *Limnopus* trackways have since been discovered in

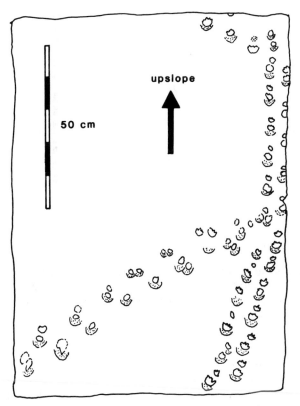

FIGURE 2.13 *Laoporus* trackways in the Lyons Sandstone that progress upslope at an oblique angle are reminiscent of trackways from the Coconino Sandstone.

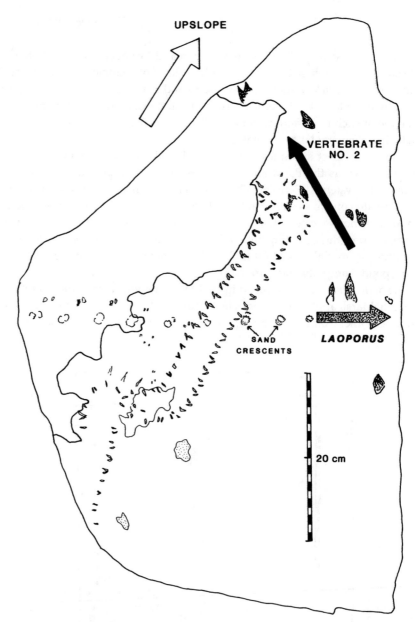

FIGURE 2.14 This map of a slab of Coconino Sandstone shows an invertebrate trail that terminates abruptly—possibly signifying predation by a vertebrate. Alternatively, the trackmaker might simply have flown away. (Compare with Figure 2.15.) *Based on a specimen in the Minnesota Science Museum.*

West Virginia and other eastern states. There are similarities between *Limnopus* and the *Baropezia* tracks we already discussed in the Carboniferous section of this chapter. *Baropezia* looks very much like *Limnopus* prints that show only the impressions of the ends of the blunt toes, without the full foot imprint. We note also that both kinds of tracks occur in sediments that represent alluvial fan environments.

Permian trackways from the Hermit Shale of Arizona also represent non-eolian environments. The Hermit Shale has produced distinctive tracks, which we identify as *Anthichnium* and *Dromopus*, respectively representing a diminutive salamander-like species and a lizard-like animal (figure 2.17). *Anthichnium* (meaning unknown) was first named in Europe in rocks similar to the Hermit Shale, but Gilmore introduced the name *Batrachichnus* (meaning amphibian track). By contrast, *Dromopus* (meaning not given by original author) has longer and pointed digit impressions and a distinctive fifth toe impression that sticks out to the side, as in modern lizard tracks. Little is known about the distribution of these tracks in western North America, though they have been studied more extensively in Permian deposits of Europe.

Regardless of the validity of the names preferred by various trackers, and notwithstanding the work that needs to be done in the future, these two distinctive track types have also been reported from similar non-eolian deposits in other parts of the western United States and Europe. Thus it is interesting to note that

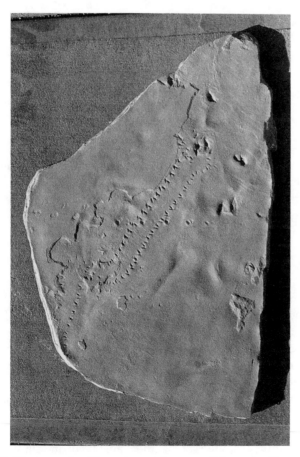

FIGURE 2.15 Photograph of large invertebrate trail from the Coconino Sandstone (see Figure 2.14 for map). *After Baird 1965.*

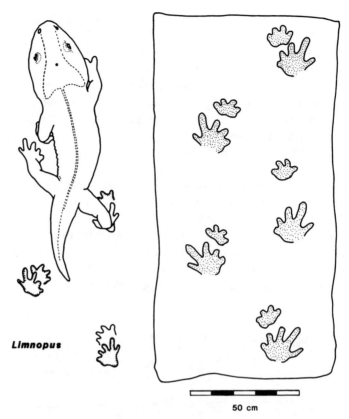

50 cm

FIGURE 2.16 *Limnopus cutlerensis* from the Permian of Colorado. Note the similarities to some of the tracks in Figure 2.5.

tracks can be used to compare animal communities from one region to another. European ichnologists have also described *Limnopus-Anthichnium* ichnofaunas that are clearly associated with amphibian trackmaking activity in swampy or humid settings. Such trace fossil assemblages suggest an entirely different ichnofacies from that just described for the *Laoporus* specimens, which are associated with desert dune environments.

All the above-described Permian tracks become particularly abundant in late Paleozoic continental deposits of North America and Europe. These geographic similarities imply the free interchange of animals across the then-supercontinent. European trackers have noted these similarities between tracks from different regions, especially in sediments of similar types and ages, and have called upon North American ichnologists to provide them with fuller descriptions of the abundant Permian tracks in the western United States. Given the large number of unstudied tracksites, North American ichnologists would do well to heed their call. Europeans have also established a useful track stratigraphy (palichnostratigraphy), which could prove useful for correlations with North America and which might ultimately lead to a refinement of estimates of the age of certain rocks. At present we can only say that there are tantalizing similarities between the Permian vertebrate ichnofacies in Europe and North America and that much more work needs to be done in the western United States.

Another Permian formation from which several tracksites have recently been reported, for the first time, is the Cedar Mesa Sandstone. One locality near Hite, Utah, has produced a particularly interesting set of footprints. The trackway, imprinted in a dune environment, shows evidence of a small animal being consumed or snatched up by a larger trackmaker (figure 2.18). The large tracks are of the *Anomalopus* type, usually attributed to primitive mammal-like reptiles known as pelycosaurs. The small tracks resemble the type known as *Stenichnus*, which, according to some authorities, are probably attributable to a primitive reptile (protorothyrid) or a small amphibian (microsaur). The trackway map shows that as the large trackmaker converged with the smaller animal, all traces

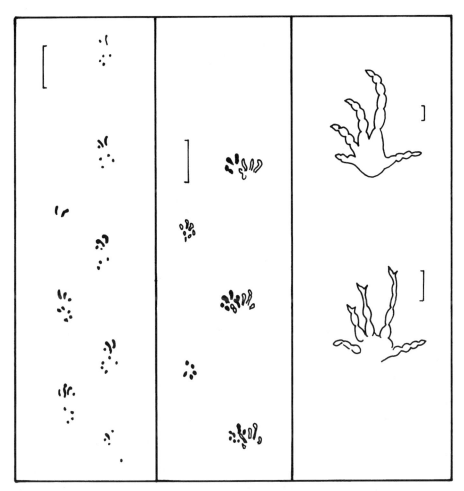

FIGURE 2.17 *Anthichnium* from the Hermit Shale of Arizona (left) was named *Batrachichnus* by Gilmore, with a similar *Anthichnium* trackway from Germany (center). Two well-preserved *Dromopus* tracks from the Howard Limestone of Kansas (upper right) and the Hermit Shale of Arizona (lower right). Scale bars are 1 cm.

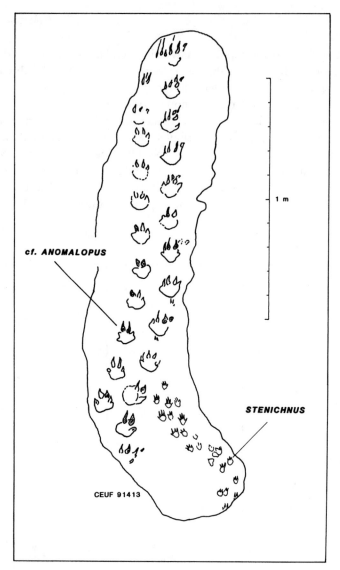

FIGURE 2.18 Trackways from the Cutler Formation, Utah, suggest a predator (cf. *Anomalopus* trackmaker) disposing of its prey (*Stenichnus* trackmaker). *After Lockley and Madsen 1992.*

of the latter disappear near the point where the two trackways intersect. This suggests that the smaller animal was snapped up and eaten or carried off.

Reports of trackways documenting attacks or "murders" of prey by predators are rare. Moreover, few such interpretations are widely accepted. But the Cedar Mesa specimen is sufficiently clear as to qualify as one of the better attack scenarios documented to date. Unfortunately, although trackway photographs and replicas are available for study, the original site is now underwater, having been flooded after the construction of Glen Canyon dam in 1973. The good news, however, is that seven more Cedar Mesa tracksites have come to our attention—suggesting no shortage of future research opportunities.

It is interesting to note that both *Anomalopus* and *Stenichnus* also occur in the Wescogame Formation of the Grand Canyon area. Although this late Carboniferous deposit is a little older than the Cedar Mesa Sandstone of the early Permian, it represents a similar eolian environment. Thus the co-occurrence of these two track types may have some ecological significance. This would not surprise us, given the growing evidence for strong relationships between trackmakers and their habitats.

The Latest Big Discovery

Undoubtedly one of the most exciting current developments in Permian track research in the western states is the Paleozoic Trackway Project, centered in the Robledo Mountains near Las Cruces, New Mexico. Tracks were first reported from the area's Abo Formation in the 1930s, but at that time the fossil footprints were believed to be mainly small and of limited scientific importance. This perception has changed dramatically in recent years, thanks to the work of Jerry MacDonald, a graduate of the New Mexico State University Earth Science Department.

Armed with the general knowledge that tracks do indeed occur in the Abo Formation, MacDonald set about searching for productive beds. Eventually, he excavated a substantial track record. His efforts to date have been extremely rewarding, resulting in discoveries that many, including MacDonald himself, have hailed as nothing short of spectacular.

Most of what the paleontological world knows about MacDonald's work has thus far come from published interviews with MacDonald and from magazine articles, rather than from scientific documentation. However, subjective as some of the reports are, MacDonald's success in the Robledo Mountains underscores the wealth of untapped Paleozoic trackway data that awaits discovery in the western United States. During a search that lasted approximately five years, MacDonald discovered over thirty tracksites, including his main excavation site where about two dozen track-bearing levels occur in sequence in only a few meters of strata.

In addition to demonstrating that tracks are abundant in the Abo Formation, MacDonald has shown that they are diverse and well-preserved. Although a full description of tracks and traces is not yet available and will take many years to compile, preliminary information indicates that the main excavation site yields a wide variety of vertebrate and invertebrate traces—ranging from the trackways of large pelycosaurs (fin-backed reptiles), through a variety of intermediate-sized and small reptiles and amphibians, and even the delicate tracks and traces of diminutive invertebrates. Some of the pelycosaur tracks (figures 2.19–2.21) may have been made by the well-known fin-backed reptile *Dimetrodon*, whose tracks have been correlated with the ichnogenus *Dimetropus* (meaning, Dimetrodon foot). *Dimetropus* has also been reported from non-eolian red beds elsewhere in North America and from localities in Europe. Following the theme of the paleoecological discussion introduced in the previous section, it is probable that *Dimetropus* trackmakers frequented non-eolian environments.

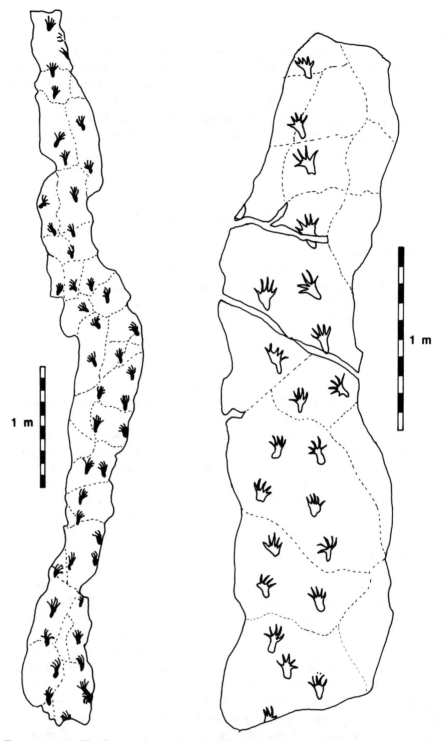

FIGURE 2.19 Trackways from the Abo Formation of New Mexico. *Left:* a 25-foot-long *Dimetropus* trackway, showing 47 consecutive footprints (in the United States National Museum, Smithsonian). *Right:* a 12-foot-long pelycosaur trackway, showing 23 tracks (in the Los Angeles County Museum). *After MacDonald 1990.*

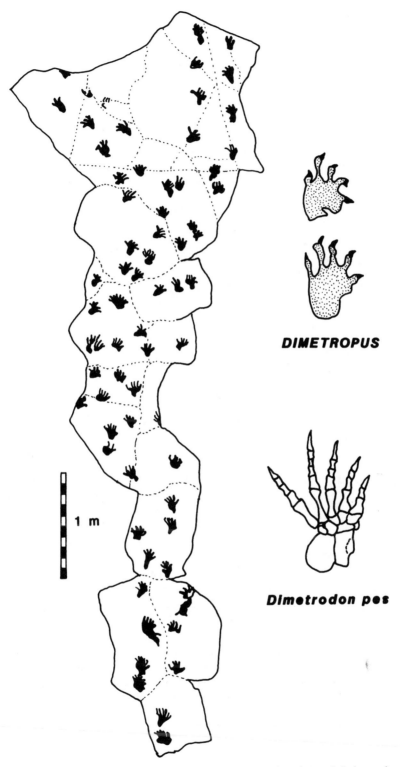

DIMETROPUS

Dimetrodon pes

FIGURE 2.20 Pelycosaur tracks (*Dimetropus*) in a 25-foot-long slab from the Abo Formation of New Mexico, showing 72 footprints (Carnegie Museum Collections). *After MacDonald 1990.*

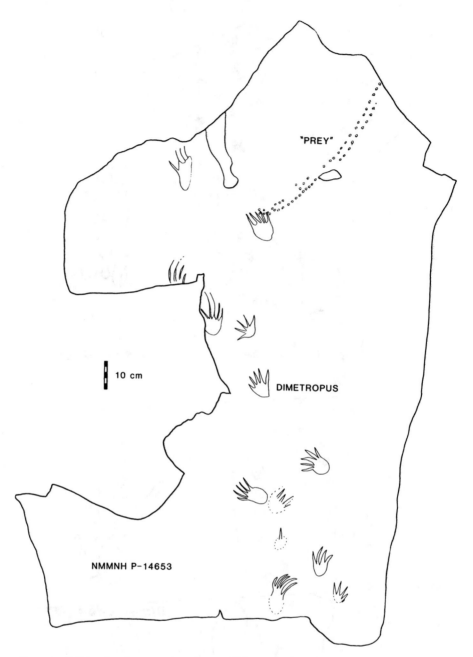

"PREY"

10 cm

DIMETROPUS

NMMNH P-14653

FIGURE 2.21 A pelycosaur trackway (*Dimetropus*) intersects the trackway of a smaller tetrapod, suggesting predation. (Compare with Figure 2.18.) Further study of this New Mexico specimen reveals that the trackway of the purported "prey" individual continues beyond the *Dimetropus* trackway, where it is faint, owing to different substrate conditions (Hunt and Lucas, 1998). These authors also question purported predation evidence from Utah and Arizona (Figs 2.14, 2.15, and 2.18).

Many of the Abo trackways are of exhibit quality. At first, MacDonald excavated large trackway segments for the Smithsonian Institution, the Carnegie Museum of Natural History, and the Los Angeles County Museum, as well as for the National Science Museum in Tokyo. In 1994, with the help of New Mexico Senator Jeff Bingaman, the Bureau of Land Management (BLM) acquired funds to initiate a project involving the Smithsonian Institution, the New Mexico Museum of Natural History and Science, and our own research group at the University of Colorado at Denver.

During the course of our initial studies we have collaborated with Jerry MacDonald to transfer all the track specimens to the New Mexico Museum of Natural History and Science. In so doing we have had the opportunity to describe another particularly interesting specimen. Like the famous Cedar Mesa tracksite in Utah (figure 2.14), this specimen (figure 2.21) also reveals evidence of Permian predation. In this case the *Dimetropus* tracks intersect the trackway of a much smaller animal whose footprints have not yet been assigned a name. At the trackway intersection all signs of the smaller animal disappear, strongly suggesting that the larger animal unceremoniously snapped it up.

Although evidence is still in short supply, such finds suggest that convincing evidence for predator-prey interaction is currently a little easier to come by in Permian sediments than in younger deposits. We can only speculate on why this should be so, and several possible explanations spring to mind. During the early Permian, reptiles such as *Dimetrodon* and its contemporaries were cold blooded and thus had smaller food requirements than warm-blooded animals of the same size. This is thought to have reduced the frequency of predation quite significantly and perhaps made potential prey species less inclined to be alert when in the vicinity of potentially dangerous predators. Moreover, these animals may not have moved so far or so fast as their warm-blooded descendants when searching for food. If this assumption is correct, then there is a greater chance of finding evidence of predator-prey encounters in a small area of Permian track-bearing surface than on a younger surface of equal size. It is also possible that cold-blooded Permian vertebrates used open areas such as mudflats to bask, instinctively aware that the possibility of high speed attacks was relatively low. Any combination of these scenarios helps account for the high concentration of tracks in Abo sedimentary deposits and the seemingly greater frequency of encounters between predator and prey within small areas.

Like many "big" finds, MacDonald's discoveries have generated controversy and much media coverage. The story deserves brief attention here, but the reader can find a detailed account in MacDonald's autobiography, *Earth's First Steps*.

After graduating in earth science from New Mexico State University, MacDonald continued his education as a graduate student in sociology at the University of Virginia but continued to live part-time in New Mexico. When he first began to find abundant tracks he approached New Mexico State University in search of research stipends. Budgetary constraints were such that no financial support was forthcoming. Apparently, some faculty questioned the significance of the discoveries and suggested that their importance was exaggerated. MacDonald next approached several large museums, where he did find allies who took his claims seriously and where he was able to obtain research support. These museums also issued press releases describing the find as "world class" and the specimens as "glorious." Thus began the Paleozoic Trackway Project.

Reports of MacDonald's discoveries began to appear in popular but reputable geology magazines, such as *Geotimes* and *Earth Science*, as well as in the more general *Science Probe* and *National Geographic*. This fostered further controversy in New Mexico, marked by newspaper headlines such as the one proclaiming "Scientist wins world acclaim, but is snubbed in New Mexico."

The story of the Paleozoic Trackway Project has several other interesting twists. Because the tracks are on public land, MacDonald's excavations are within the jurisdiction of the BLM, and all specimens technically belong to the Smithsonian Institution unless another authorized repository is approved (the New Mexico Museum of Natural History and Science, in this case). As large trackway segments—up to 24 feet in length—were extracted, storage and security became a problem. One 17-foot trackway was stolen from the site immediately before the director of the New Mexico office of the BLM was scheduled to visit it. BLM officials then offered a $1,000 reward for the stolen property, and the tracks were returned within two weeks.

At the time of our writing, Jerry MacDonald's book is in press and the official Paleozoic Trackway Project is barely under way. But we already know enough to conclude that the Permian and late Paleozoic constitute the new frontier in trackway studies. The prospects for further significant discoveries are indeed promising. Although it is hard to predict exactly where these discoveries will lead, it seems that, in a general sense, they will lead where track discoveries from other time periods have led—to a substantial new data base and to the realization that tracks are far more common than bones in late Paleozoic continental deposits of the western United States.

Our sense that we are now participating in a fundamental and fascinating episode of vertebrate paleontology helps set the stage for our description and interpretation of the rest of the track record. As a general rule, the further we look back in time, the more difficult the track record is to decipher. On the one hand this helps explain why the Paleozoic track record is still rather poorly understood—though lack of study is also a contributing factor—but, on the other, it leads us to expect that the track record of subsequent epochs will be just as rich, exciting, and perhaps even a little easier to interpret.

Triassic archosaurs: a quadrupedal rauisuchid and a bipedal lagusuchid.

3

ARCHOSAUR ASCENDANCY: THE TRIASSIC

We ought to investigate all sorts of animals
because all of them will reveal something of
nature and something of beauty.

ARISTOTLE, *THE PARTS OF ANIMALS*

■ The Triassic period marked the beginning of a new era in animal evolution and one in which all sorts of new animals came on the scene—notably the archosaurs, advanced reptiles that included dinosaurs, pterosaurs, and crocodiles. The new opportunities of this new Mesozoic era came about in part because of a major extinction that marked the end of the Permian period and the close of the Paleozoic era. Many groups of organisms were seriously affected—notably, marine creatures. Sea level dropped dramatically, exposing the continental shelves and decimating marine life. Reptiles, however, were among the groups least seriously affected. They had already developed the ability to reproduce and survive in dry continental environments. Mammal-like reptiles (also referred to as reptile-like mammals; the exact classification is controversial) were at the height of their ascendancy. The highly successful archosaurs, a group that would later include the dinosaurs, were just emerging.

At this time most of the world's continents were joined together in a supercontinent known as Pangaea II (meaning, "all world" or "wide world"). Figure 3.1 shows this massive gathering of continents. Organisms living in the interiors of large continents experience harsh "continental" climates, marked by aridity and extremes of heat and cold, unbuffered by the moderating influence of the ocean's moisture and mild temperatures. The very end of the Paleozoic was thus a time of stress and crisis induced by unusual and very severe environmental conditions, and reptiles were relatively well-equipped to cope with the demands of the new era.

The Mesozoic era began about 245 million years ago, with the Triassic period. Although the Mesozoic is often referred to as "the age of dinosaurs," the first true dinosaurs did not appear on the scene until middle to late Triassic times, at least 15–20 million years later.

Both the Permian and the Triassic fossil footprint records are extensive, and only a few specialists worldwide have acquired a deep understanding of them. Together, they reflect the complex turnover of reptilian faunas during the transition from the old Paleozoic era to the new Mesozoic era.

The term Triassic reflects the "three-fold" division of the European Triassic rock system into a lower, middle, and upper series, with only the middle representing a marine episode. In the Colorado Plateau region there is no such obvious middle marine interlude, and the middle series is largely missing. Therefore the strata fall mainly into an upper and lower series—equating to the early and late Triassic, which both represent times of terrestrial sediment accumulation. Although Triassic tracks do include footprints attributable to mammal-like reptiles, these are uncommon in North America. Instead, most of the tracks are attributable to reptiles related to lizard ancestors and, more significantly, to the even more successful group known as archosaurs.

During the Triassic, archosaurs rose from their humble origins to achieve global dominance. They were characteristic of what some paleontologists have described as a new grade of reptiles (neotetrapods = "new tetrapods") that were more advanced than were their predecessors (paleotetrapods = "old tetrapods"). Neotetrapods are distinguished by improved locomotion, including very erect posture and even, in some lineages, bipedalism. Among the many new archosaurs that arose at this time were aquatic, terrestrial, and flying forms. Other diverse adaptations include the development of armor.

Some researchers have suggested that these diverse adaptations were the result of intense competition. Others contest this view and claim that archo-

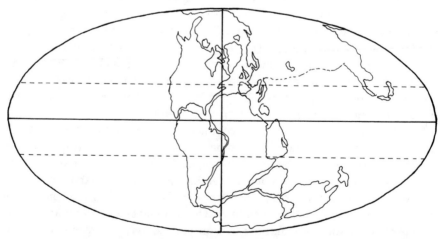

FIGURE 3.1 Paleogeographic map of the supercontinent Pangaea II in early Mesozoic time, showing all the continents joined together.

saurian success was stimulated by changes in the environment that happened to favor archosaurs over other groups. In any event, the fossil bone and track record shows that archosaurs rose to prominence and that they were among the first to respond to the challenges of the new era.

An Overview of Triassic Trackmakers and Tracks

In the sections that follow we will discuss tracks of three great branches of reptiles then extant: the mammal-like reptiles, the ancestors of modern lizards, and the archosaurs. The archosaur group was anatomically distinguished by several features of the skull and teeth but also by upright limb posture. Their trackways in turn are distinguished from the other two branches of reptiles by being narrow. Of all three groups, it was the archosaurs that became diverse to an extreme, and they are thus the principal subjects of this chapter. Some of the important groups of Triassic trackmaking archosaurs are:

Phytosaurs:	large crocodile-like aquatic forms
Aetosaurs:	large armored herbivores
Rauisuchids:	large quadrupedal carnivores
Ornithosuchids:	semi-bipedal carnivores
Ornithischians:	the first bird-hipped dinosaurs
Saurischians:	lizard-hipped dinosaurs mainly including bipedal carnivores but also including prosauropods (ancestors of the better-known sauropods or brontosaurs).

Some groups of archosaurs, such as the winged pterosaurs, are not shown in this list because their tracks are yet unknown in the Triassic. The phytosaurs are the only truly aquatic archosaurs shown in this list (they are crocodile-like, but true crocodiles are first seen in the fossil record in the Jurassic). The remaining groups in this list are fully terrestrial. The two major lines of dinosaurs are represented by the ornithischians (the bird-hipped dinosaurs) and the saurischians (the lizard-hipped dinosaurs). Although both lines are represented in the Triassic, it was not until the Jurassic that they would become the dominant reptiles.

Non-dinosaurian archosaurs were responsible for making some of the most famous and most abundant of all Triassic footprints. Of these, the tracks known as *Chirotherium* ("hand animal") are the most abundant. As shown in figure 3.2, these medium-sized tracks have five-digit impressions, most clearly seen on the hind feet, making them remarkably similar in size and appearance to the imprint of a human hand. The thumb-like appendage, however, is deceiving. It is not a true thumb (our digit I); instead it is the outside digit (digit V), equivalent to our little finger or, because it is a hind foot, our little toe. After the original discovery of *Chirotherium* in Europe in 1835, paleontologists were quite unable to suggest which animal group had been the trackmaker. The so-called thumb impression on the hind foot caused so much confusion that even eminent paleontologists reconstructed the trackmaker as a grotesque animal that literally crossed its right foot over beyond its left and its left foot over beyond its right as

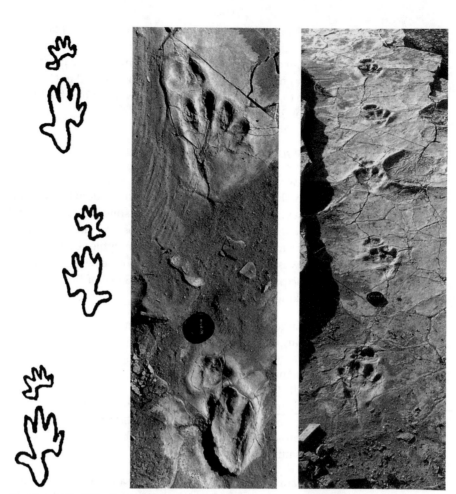

FIGURE 3.2 *Chirotherium* footprints (*left*), made by archosaurian reptiles, were first described in the 1830s. Note the narrow trackway pattern and the hand-shaped appearance of hind footprint. Photograph (*center*) shows a *Brachychirotherium* trackway from the Sloan Canyon Formation, northeastern New Mexico, with detail of the two manus-pes sets (*right*).

it walked (figure 3.3). A moment's reflection suggests that this is a highly ineffi-cient and unlikely gait.

It was not until the twentieth century that the mystery of the *Chirotherium* trackmaker was solved. First, skeletal remains of early Triassic archosaurs had to become better known. A South African specimen of a small archosaur named *Euparkeria* was shown to have a foot that nicely fit the *Chirotherium* footprint pat-tern, even though the animal was too small to explain medium- and large-sized *Chirotherium* tracks. More important, the pattern of footprints within *Chirotherium* trackways was reinterpreted in a credible way that did not make the trackmaker ungainly and inefficient. Later, in the 1960s, certain *Chirotherium* track types were linked with larger archosaurs like *Ticinosuchus*, one of several quadrupedal, car-nivorous, rauisuchids that were the "lions" of their time (figure 3.3).

Today we regard rauisuchid forms like *Tichinosuchus* as just one of a large number of archosaur groups that left *Chirotherium*-like tracks. We find footprints of this general type (and in a wide range of sizes) in much of the Triassic rock record throughout most of the world. In general, the various kinds of *Chirotherium*-like tracks have been attributed to archosaurs including rauisuchids, aetosaurs, phytosaurs, ornithosuchids, and several lesser-known groups.

Chirotherium-like tracks, particularly those of the hind footprints, indicate that much of the animal's weight was carried on the three central digits. In this respect the tracks can be viewed as forerunners of the many types of three-toed dinosaur tracks that become abundant by the end of the Triassic period.

The Moenkopi Formation of the Early and Middle Triassic

Only one deposit in the Colorado Plateau region is well known for producing early and middle Triassic tracks. This is the Moenkopi Formation, named for an Indian village in northern Arizona. In this locality it is usually divided into three members—Wupatki, Moqui, and Holbrook. The upper two members, the Moqui and the Holbrook, have yielded tracks. The Moenkopi Formation also outcrops to the north in Colorado and Utah and to the east in New Mexico. Skeletal

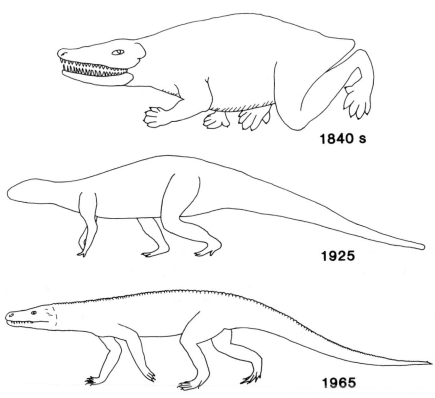

FIGURE 3.3 The first and incorrect reconstruction of the *Chirotherium* trackmaker, with a later and a modern reconstruction for comparison. Note that the old reconstruction has the animal crossing its feet in an awkward manner. *After Haubold 1984.*

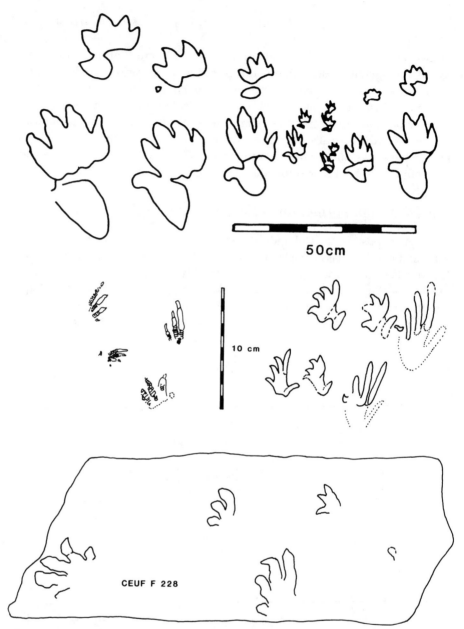

FIGURE 3.4 *Chirotherium* tracks (top) from the Triassic Moenkopi Formation of Arizona show a wide range of sizes and indicate a variety of quadrupedal archosaurs. *Rotodactylus* tracks (middle left) with skin impressions (after Peabody 1948). Lacertoid footprints, *Akropus*, (middle right) from the Lower Moenkopi of Arizona (after Peabody 1948). *Akropus*-like tracks (bottom) from the Moenkopi of Utah (based on College of Eastern Utah specimen CEUF F 228).

50cm

10 cm

CEUF F 228

remains from the Moenkopi Formation are primarily those of amphibians. Tracks, however, are mainly those of archosaurs and other reptiles, and these are known from all four states.

This obvious inconsistency between the track record and bone record is not uncommon. It reflects the fact that many sediments accumulate in aquatic environments, such as ponds, lakes, and river channels, thus burying mainly aquatic animals. By contrast, tracks made along the shorelines of these bodies of water reflect the comings and goings of the terrestrial fauna. Such evidence alerts us to the importance of tracks in sampling parts of the ancient animal community that are poorly represented or missing in the bone record.

Most of the track-bearing beds of the Moenkopi record times rather late in the early Triassic or rather early in the middle Triassic. The most common tracks appear to have a *Chirotherium* affinity. These are often associated with various other reptilian footprints (figure 3.4). One distinctive track type found in the early Triassic of Arizona and Utah is known as *Rotodactylus*, referring to a digit impression that is rotated backward. This particular track type is interesting because a controversial suggestion has been made correlating the tracks with the bones of a presumed dinosaur ancestor, *Lagosuchus* (the rabbit crocodile), or one of its relatives. We are skeptical about this interpretation and prefer to follow other trackers who are more inclined to infer that the *Rotodactylus* trackmaker was probably some kind of lizard relative (lepidosaur). In general, both the foot structure and the pattern of scale impressions seen on some specimens provide support for this latter interpretation. Other lizard-like (lacertoid) tracks referred to as *Akropus* or *Rhynchosauroides* also occur in these beds (figure 3.4).

A number of *Chirotherium* tracks and other reptilian tracks from the Triassic show distinctive skin or scale impressions. Frank Peabody conducted extensive studies of early Triassic tracks in the 1940s and 1950s and published classic

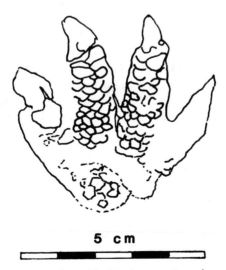

5 cm

FIGURE 3.5 *Chirotherium* tracks with skin impressions from Wupatki Crater area of Arizona.   *After Peabody 1948.*

works. His illustrations of *Chirotherium* tracks with skin impressions are among the best ever published. An excellent example is shown in figure 3.5. It is a track from the Moenkopi Formation at Wupatki Crater. Replicas of this specimen are available for purchase at the Museum of Northern Arizona.

Among Peabody's other noteworthy accomplishments in the field of ichnology was his pioneering use of modern animals in trackmaking experiments. This allowed him to more confidently identify the affinity of ancient trackmakers through a better understanding of their track and trackway morphology. Peabody also tried to establish guidelines for the quality of tracks needed in order to establish new ichnological names. He rightly pointed out that trackers need to find at least three consecutive tracks in a trackway before they can begin to interpret the gait of the trackmaker. Unfortunately, not all trackers have adhered to this common sense rule. Especially in earlier times new, and often dubious, names were proposed on the basis of very poor trackway material.

In addition to studying many different types of normal *Chirotherium* tracks made by walking animals, Peabody also studied some unusual tracks from the Moenkopi Formation in the vicinity of Dinosaur National Monument. These tracks appear to have been made by large animals that were either swimming or slipping and sliding on a very soft substrate. In either case these tracks are very irregular and difficult to decipher. In several instances the tracks are parallel, slightly curved, and elongate. These "scrape" marks, in some cases, resemble large turtle tracks that have been described from Jurassic deposits in France. Examples of these unusual tracks are held at the Utah Field House Museum in Vernal, Utah, and in the joint collection of the University of Colorado at Denver and the Museum of Western Colorado. Additional examples of "swim tracks" in red beds of the same Triassic age have been reported from Wyoming.

Recently, Moenkopi tracks that include both walking and "swim traces" have been discovered near Hite on the shores of Lake Powell. Unlike the excellent preservation seen in some walking tracks that may even include skin impressions, the swim traces are poorly and irregularly preserved. Although poor in quality these irregular swim traces cannot be ignored, as they are now known from a large number of sites. Since the tracks were evidently made in a shallow aquatic environment, it is possible to speculate that the trackmakers were amphibians. But as pointed out earlier, tracks of aquatic animals are rare and are sometimes preserved in unusual configurations that indicate actual swimming behavior. Recognizing such abundant swim tracks helps explain the inconsistency noted earlier. These are evidently the traces of the amphibians that dominate the bone record at most sites.

Such unusual or irregular tracks are among the most controversial of all fossil footprints found in any geologic time period. They are often mistakenly attributed to the wrong trackmakers (see chapter 5) or interpreted as unusual behavior such as swimming, hopping, or erratic changes in direction. Such tracks require careful study to distinguish between behavioral and preservational effects and thereby avoid serious misinterpretation. Examples of some of the controversies that have arisen are discussed later, in reference to other problematic footmarks.

The final example of tracks from the Moenkopi Formation pertains to some recently described trackways from the town of Holbrook, Arizona. These include what appears to be the first example of mammal-like reptile trackways found in the Moenkopi Formation. The trackways (figure 3.6), which are preserved in rocks of middle Triassic age, have been assigned the new name *Therapsipus* (meaning, footprints of a therapsid), and it was clearly made by a large, wide-bodied animal. In the early and middle Triassic, the therapsid group of mammal-like reptiles is indeed the best candidate for making this type of trackway. The Holbrook discovery thus highlights the utility of tracks in revealing the presence of animal groups that are not known in the bone record in this region.

Tracks Around the World

It is very interesting to compare the distribution of bones and tracks of different types of reptiles and amphibians in different parts of the world during the early and middle Triassic. In the Moenkopi Formation, and in similarly aged rocks in parts of Europe, amphibians are the most common animals indicated by fossil bones. But according to the track record of both regions, the archosaurs (*Chirotherium* trackmakers) were the most abundant, with a lesser number of other reptiles (therapsids and lepidosaurs) also present. In other regions, especially southern Africa, India, and China, the skeletal record is dominated by therapsids. Thus the track and bone records seem to sample quite different animal communities,

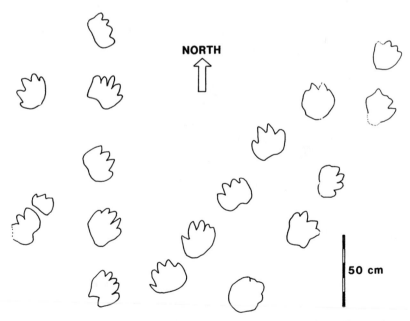

FIGURE 3.6 Recently discovered trackways from the Moenkopi Formation at Holbrook, Arizona, have been named *Therapsipus* and attributed to large, mammal-like reptiles. *After Hunt et al. 1993.*

and the bone record itself can be substantially different depending on which fragment of the old Pangaea II one surveys. Surely, bone and track records are *both* essential for a complete picture of vertebrate life.

Another example of controversy over tracks was introduced in chapter 2 in our discussion of horseshoe-like markings, of uncertain origin, found in Paleozoic rocks of the Grand Canyon region. Similar horseshoe-shaped markings have been observed in the Mesozoic Moenkopi Formation (figure 3.7). These particular tracks were studied by Frank Peabody, who identified them as "current crescents" caused by the scour of currents around a small pebble or other obstruction on the sediment surface. Peabody was aware of claims that such features, also found in Triassic rocks in Germany and in Jurassic rocks in the northeastern United States, were of vertebrate origin, and he was anxious to make a correct interpretation for the Moenkopi occurrences. We ourselves have seen such features in Moenkopi sediments and agree with Peabody that they are current crescents, not tracks. There are, however, other horseshoe-shaped markings found in other sedimentary rocks elsewhere that are not current crescents.

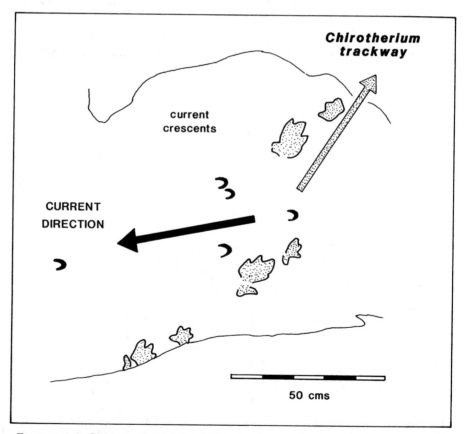

FIGURE 3.7 Current crescents and *Chirotherium* tracks from the Moenkopi Formation.   *After Peabody 1947; Lockley et al. 1993.*

Which Came First, the Dinosaur or the Track?

The suggestion that dinosaur tracks occur in early or middle Triassic rocks, thus before the first record of dinosaur skeletons (in the late Triassic) is controversial—not least because it implies that tracks may be more useful than bones in dating the origin of certain important groups.

The fossil bones of late Triassic dinosaurs from South America, including *Eoraptor*, *Herrarasaurus*, and *Pisanosaurus*, have recently received a lot of publicity as "the world's oldest dinosaurs." Dinosaur skeletal fossils that are at least as old, or about the same age, have been documented in late Triassic deposits in India, Morocco, and North America. Some of these dinosaurs are very primitive, but others are not. The more evolved or derived types suggest the possibility of a significant period of dinosaur evolution prior to their appearance in the rock record of the late Triassic. Such an argument could be used to support the interpretation that dinosauroid, or "dinosaur-like," tracks of the early and middle Triassic really were made by true ancestral dinosaurs. The tracks, not the skeletal fossils, would thus have to be treated as the earliest evidence of dinosaurs. Recent work however suggests that certain claims of pre-late Triassic dinosaur tracks are invalid and based on poor material. But whether the earliest record is a track or a skeleton, it is certain that the oldest evidence is unlikely to represent the actual oldest dinosaur type.

Overall, paleontologists expect that as the years pass new finds will push back the origin of any particular group of animals. Such evidence can include tracks. For example, the oldest tetrapod tracks are evidently older than the oldest tetrapod skeletal remains. Similarly, a case can be made that the oldest hominid tracks, from the famous Laetoli site in Tanzania, are slightly older that the oldest skeletal remains. We should not be surprised, then, if the dinosaur record presents a similar sequence.

As so few North American tracks of middle Triassic age have been studied in detail, most of what has been reported about important changes in the ancient track record comes from studies by George Demathieu of middle Triassic tracks from France. Among the tracks reported from this region are small, three- and four-toed tracks that appear to document the emergence of an advanced lineage of archosaurs. The advanced foot structure and gait indicated by these tracks strongly suggests that the trackmakers were related to early dinosaurs. Dinosaur fossils have not been found in rocks this ancient anywhere else in the world. Although we cannot say for certain whether or not these tracks were actually made by true dinosaurs, or "dinosauroids" in the terminology of Demathieu, they do indicate the appearance of reptiles that were very much like dinosaurs.

In South America, middle Triassic rocks have yielded the fossil bones of *Lagosuchus*, the so-called rabbit crocodile. *Lagosuchus*, an early archosaur, is a popular candidate for a dinosaur ancestor. This animal and its dinosaur-like relatives have not yet been convincingly linked with a particular track type, so it is entirely possible that *Lagosuchus* tracks, or at least tracks of lagosuchid-like species, may be among those reported from the middle Triassic of France. Because of the good track record of this epoch in France, Demathieu and his colleagues have been able to study this important phase of early archosaur evolution in some detail. Among their conclusions is the strong inference that the best record of the earliest phases of archosaur evolution and initial diversification is currently found in the track record, and that the transition from pre-dinosaurian archosaurs to true dinosaurs can be tracked in the footprint record.

The lack of an extensive middle Triassic sedimentary record in North America, though lamentable, nevertheless offers an instructive lesson. If we were to compare the early, middle, and late Triassic track records without being aware of the missing strata, we would be struck by the sudden appearance of advanced archosaurs, including bipedal dinosaurs, without intermediate forms. This sudden appearance in the rock record could then be misinterpreted as a sudden jump in the actual evolution of archosaurian locomotion and posture. Thankfully, the middle Triassic is well recorded in other parts of the world, where it displays a more expected progression of archosaurian evolution.

The Chinle Group of the Late Triassic

Late Triassic tracks are known from terrestrial deposits at several dozen localities in Colorado, Utah, Arizona, New Mexico, Oklahoma, and Wyoming. Although most of these geological deposits are known mainly to specialists, at least one—the Chinle—is famous for the abundant fossil logs it bears in the Petrified Forest of Arizona. Its brightly colored sediments also form the nearby Painted Desert. Traditionally, the widespread late Triassic deposits of the western states were classified by geologists into dozens of different formations and members. According to recent publications this scheme can be simplified; all these units can be considered part of the Chinle Group. For the purposes of this discussion we have distinguished Chinle track assemblages in three different geographical areas:

- *Southern Colorado Plateau Region* (the Chinle Group of northern Arizona and New Mexico and southern Utah and Colorado)
- *Eastern Region* (the Chinle Group of eastern New Mexico, Oklahoma, Texas, and Colorado; formerly the Dockum Group)
- *Northern Colorado Plateau Region* (the Chinle Group of northern Colorado and Utah and southern Wyoming in the environs of Dinosaur National Monument)

These three regional representations of Chinle Group deposits in the western states have been extensively studied, revealing many kinds of plant and animal fossils, including dinosaurs and a variety of other reptiles. The deposits themselves have also been subdivided into a series of distinctive sedimentary units representing ancient river bank, flood plain, or lake deposits. Some of these habitats were ideal for track formation and preservation. Most of the tracks that have been found in Chinle strata were made on the shorelines of ancient waterways. The late Triassic was generally a time of distinctly seasonal precipitation in western North America, and some geologists have dubbed this epoch the time of "the megamonsoon." Abundant evidence of wet conditions is seen in the fossils of various lake- and river-dwelling snails, clams, crustaceans, fish, and amphibians. While the traces of some of these creatures are known, it was the land-based reptiles that left the majority of footprints. High seasonal rainfall, resulting in seasonally high water tables, made for excellent conditions of track formation and preservation.

THE SOUTHERN COLORADO PLATEAU REGION OF THE CHINLE

Several tracksites are known in the Southern Colorado Plateau Region of the Chinle Group. One, however, stands out beyond all others because of its areal extent and the variety of tracks it offers. This site is known as the Shay Canyon locality, which is northwest of Monticello in eastern Utah. The track-bearing layer represents the activity of several dozen animals that lived at a single geologic instant late in Chinle time.

The Shay Canyon site reveals over 250 footprints on the surface of a single layer (figure 3.8). These tracks are mainly of a particular variety of *Chirotherium* known as *Brachychirotherium* ("broad" Chirotherium; figure 3.9) and were probably made by the large herbivorous armored archosaurs known as aetosaurs. Other tracks from the site include a large *Atreipus*-like form (figure 3.10) that is different from the classic *Atreipus*, which will be described later in this section. The *Atreipus*-like track clearly qualifies as one of the largest three-toed dinosaur tracks known from the late Triassic. Most of the other footprints from Shay Canyon are problematical. A few closely resemble *Pentasauropus*, which several trackers infer to have been the tracks of large mammal-like reptiles known as dicynodonts (figure 3.10). There are also a number of enigmatic scratch and scrape marks at this site and some scattered coprolites (fossil dung). The scrape marks were at first interpreted as swim tracks made by phytosaurs, but we regard this suggestion as speculative.

One other Chinle Group tracksite is worth mention. Located in the valley of the Dolores River near Gateway, Colorado, the Alabaster Box site has yielded an assortment of small, three-toed tracks and other footprints (figure 3.11, inset). The small, three-toed tracks are typical of the footprints usually referred to as *Grallator*, and which are usually correlated with early dinosaurs like *Rioarribasaurus* (formerly named *Coelophysis*).

Also shown in figure 3.11 is the partial trackway of a relatively small quadrupedal animal, with a fairly wide trackway. No one can be certain of the affinity of this trackway. The tracks occur in sandstones that were deposited in a small river system that flooded its banks.

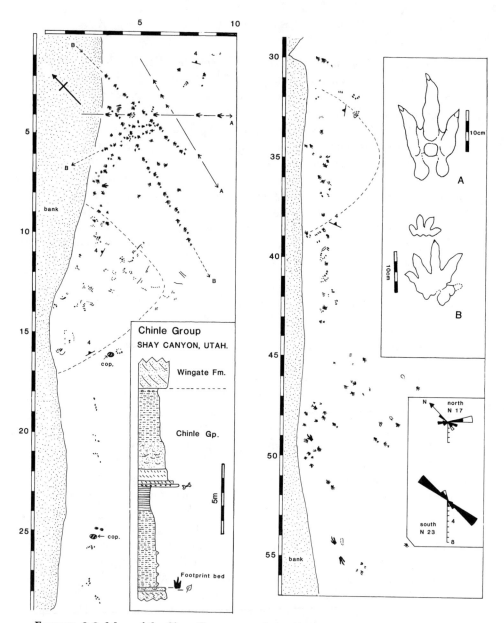

FIGURE 3.8 Map of the Shay Canyon tracksite ("cop." = coprolite). *Insets:* stratigraphic position of track-bearing bed, trackway orientations for the northern and southern parts of the site, and details of a theropod track (A) and a *Brachychirotherium* manus-pes set (B).

Tracks of crayfish-like arthropods are also found at this site, and it is tempting to view these arthropods as part of the food supply of the carnivorous, three-toed theropod dinosaur that made the *Grallator* tracks. Large vertical burrows at the Dolores Valley site were at first interpreted as those of lungfish, in accordance with popular opinion in the 1980s. Recent studies suggest, however, that these burrows may instead be attributable to crayfish. The bird- or even heron-like

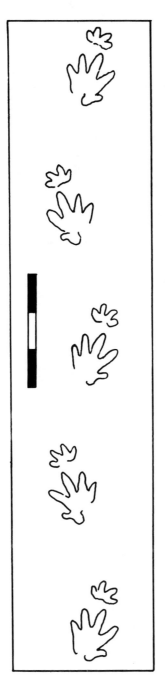

FIGURE 3.9 *Brachychirotherium* trackway from the Chinle Group, Utah, based in part on CU-MWC 151.1–151.14. Scale bar 30 cm.   *After Lockley 1986.*

design of *Rioarribasaurus* and its relatives would no doubt have equipped this theropod for foraging in the shallows of rivers, lakes, and ponds. Surely these dinosaurs would have taken advantage of a food supply that included crayfish.

One of the most recent footprint discoveries in the Chinle Group happened at a site on the northern shores of Lake Powell. This site has yielded tracks known as *Atreipus milfordensis* (named after the nineteenth-century tracker Atreius Wanner and the town of Milford, New Jersey, where these tracks were first found). These tracks were previously known only from eastern North America and a single locality in Europe. Thus the Lake Powell discovery is the first report

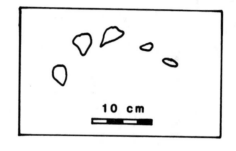

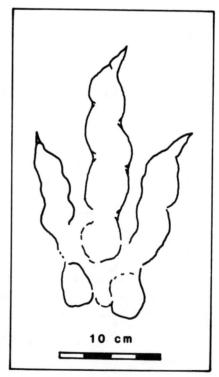

FIGURE 3.10 *Top:* Pentasauropus *track from the Chinle Group, Shay Canyon, Utah.   Based on CU-MWC 152.1–152.3. Bottom: Large Atreipus-like footprint. Based on CU-MWC 150.1–150.6.*

The Coelophysis Crisis

Coelophysis is one of the best known of all Triassic dinosaurs. A few years ago this theropod was adopted as the "state fossil" of New Mexico. The incomplete remains of the theropod dinosaur originally named *Coelophysis*, however, pose various problems.

The original specimen from which *Coelophysis* was described and named (in 1887, by the famous nineteenth-century dinosaur hunter Edward Cope) was based on very poor material excavated from Rio Arriba County, New Mexico. When subsequent discoveries of better theropod dinosaur material were made at different localities in the same area, the name *Coleophysis* was applied to these new finds without any proof that the new material represented the same dinosaur genus found by Cope. Among these new discoveries of theropod dinosaurs were the spectacular finds at what has become the Ghost Ranch quarry. This site is several miles from the original discovery site, and it was found by paleontologists from the American Museum of Natural History in 1947. The Ghost Ranch quarry has since yielded dozens of complete or virtually complete specimens, widely referred to as *Coelophysis*.

According to the International Code of Zoological Nomenclature, a fossil name may be "suppressed" or declared "invalid" if the original material is shown to be inadequate, or if a known specimen is given an additional or redundant name—as in the famous case of the name *Brontosaurus*, which was given to a dinosaur that had already been named *Apatosaurus*. The I.C.Z.N. rules also allow paleontologists to establish or "erect" new names for better material that can be shown to be unrelated to the original problematic material. This is what has happened with *Coelophysis*. Most professionals agree that the original specimen is a poorly preserved or scrappy fossil, with no diagnostic features that distinguish it from other theropods. Therefore by the I.C.Z.N. rules, *Coelophysis* is a dubious name, and one that can only apply to the original specimen. It would be inappropriate to use such a name for other specimens.

For this reason, one of us (Adrian Hunt) in a paper coauthored with Spencer Lucas recently proposed that a new name, *Rioarribasaurus*, apply to the good, new material from the Ghost Ranch Quarry. They argued that this material has the diagnostic features necessary to name a new dinosaur species. Although the Hunt and Lucas proposal was not particularly bold according to the standard academic approach for the renaming of a fossil, the proposal created quite a stir, mainly because the name *Coelophysis* is quite well known and widely used. At the time of our first writing this book, various

(continued)

petitions and counter-petitions had been presented to the I.C.Z.N., and the future of the name "*Coelophysis*" was uncertain.

The debate was decided in favor of retaining the name *Coelophysis*, for "stability" of nomenclature. However, it was recognized that the type material of Coelophysis was so poor that a new type specimen (neotype) was designated (the one that Hunt and Lucas had wanted to call *Rioarribosaurus*). So trackers will still talk about *Coelophysis* (or *Coelophysis*-like) tracks, as noted below. We should note however that *Coelophysis* is not a coelurosaur, even though it was so designated in older literature. Technically the concept of coelurosaur tracks is invalid, by definition for the Triassic, because coelurosaurs had not yet evolved.

Another complication in the use of terminology for theropod trackmakers stems from the old tradition of referring to tracks of large, robust theropods as "carnosaur" tracks and those of smaller, slender or gracile animals "coelurosaur" tracks. "*Coelophysis*"-like dinosaurs were often referred to as coelurosaurs. This simple binary classification of theropods is no longer widely accepted. Recent work on theropod relationships provides new definitions. Two main lineages are still recognized, but these are now called the ceratosaurs and the tetanurines, which diverged from one another very early in theropod history. The tetanurines can be divided into the Carnosauria and Coelurosauria. The latter group, which includes birds, arose sometime after the Triassic. Because there is technically no such thing as a Triassic coelurosaur track, footprints made by *Coelophysis* and its close relatives must now be referred to as ceratosaur tracks or simply as theropod tracks.

of this footprint type from anywhere in the western states. This type of track (figure 3.12) is both distinctive and controversial. Distinctive, because its somewhat *Grallator*-like hind footprint—which is typical of small bipedal theropod dinosaurs, except for the lack of sharp claw impressions—is combined with small, tulip-shaped front foot (or manus) impressions that show that the trackmaker typically went around on all fours. The *Atreipus* track type is also controversial because some trackers have attributed it to a small theropod and others to a small ornithopod. In a recent comprehensive study Paul Olsen and Don Baird opted for the ornithopod interpretation. If their interpretation is correct, then *Atreipus* represents the earliest known ornithopod trackway. If not, the evidence would seem to suggest that early theropods could and did walk on all fours, a behavior that is virtually unknown in the established theropod track record. Either way, therefore, *Atreipus* appears destined to challenge our understanding of early dinosaurian trackmakers.

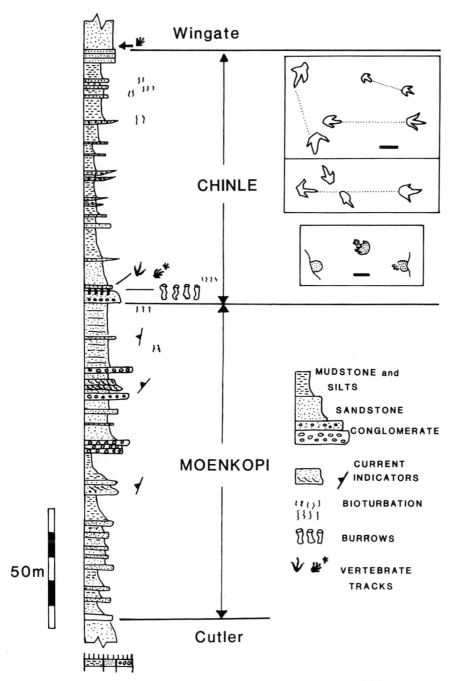

FIGURE 3.11 Three-toed tracks made by a bipedal, *"Coelophysis"*-like dinosaurs, and five-toed trackway of a quadruped; Dolores Valley section, Colorado.   *After Lockley and Jennings 1987.*

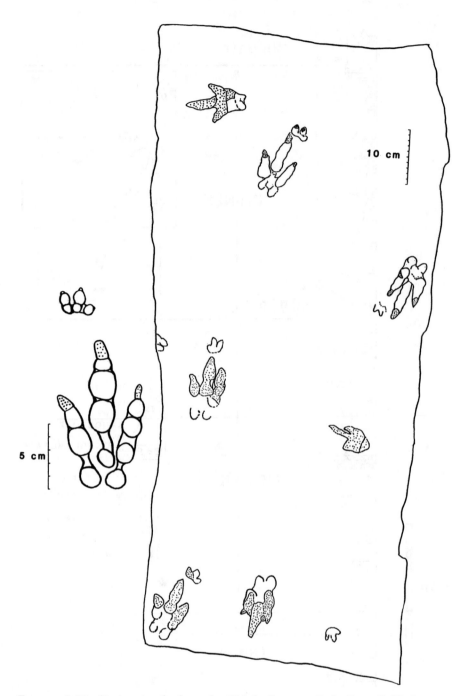

FIGURE 3.12 *Atreipus* tracks from the Chinle Group at Lake Powell, Utah. Stippled areas indicate where tracks are partially filled with sandstone. *After Lockley et al. 1992.*

Atreipus is considered by some workers to be an important track type for biostratigraphical studies (see chapter 4 for more on this application). Hartmut Haubold, a leading German ichnologist, concludes that a *Brachychirotherium-Atreipus* track zone corresponds to most of the late Triassic. We agree that both these tracks occur in the late Triassic, but we also note that recent studies suggest a slightly different picture, at least in the western United States. Specifically, we note that while *Brachychirotherium* is common in the upper (later) part of the Chinle Group, it is not known in the earlier strata. Moreover, *Atreipus* is known only from the Lake Powell locality, evidently in the middle of the Chinle Group. Thus we conclude that the track zones proposed by Haubold require some refinement. This is a subject that we develop in chapter 4.

It should be noted that *Atreipus milfordensis* (found at the Lake Powell site of the Chinle and in the eastern states) is different from the *Atreipus*-like tracks reported from Shay Canyon (figure 3.10). The Shay Canyon footprints are much larger than those of *A. milfordensis*, and they lack manus tracks. The prominent claw impressions at Shay Canyon are unusual too in that they are reminiscent of theropod tracks. The most obvious similarities pertain to the shape of the heel pad impressions.

THE EASTERN REGION OF THE CHINLE

The largest and most important site of Chinle tracks in what we call the Eastern Region (formerly called the Dockum Group) is in the vicinity of Peacock Canyon in the Dry Cimarron River valley of northern New Mexico. Like Shay Canyon, Peacock Canyon has abundant *Brachychirotherium* tracks. Peacock Canyon, however, has a greater variety of other footprint types as well. The following list gives the technical track names and the inferred trackmakers:

COMMON TRACKS
Brachychirotherium - armored aetosaurs
Tetrasauropus (or *Chirotherium* sp.) - prosauropod or other large archosaur of uncertain affinity.
Rhynchosauroides - lizard-like forms

RARE TRACKS
Pseudotetrasauropus - a small prosauropod or other archosaur
"*Coelurosaurichnus*"-*Grallator* - small bipedal dinosaur
Apatopus(?) - trilophosaurian reptile (primitive archosaur)
Therapsid track - mammal-like reptile

The Peacock Canyon tracksite is significant for a number of reasons. Not only is this the largest and most diverse assemblage of Triassic footprints known from a single site in western North America, it is also the site where several of the above-listed footprints were first reported in western North America. As such, the site is crucial to our understanding, as it provides the key for recognizing these various tracks throughout this region. The large *Tetrasauropus*-like trackways (figure 3.13) are one such important occurrence. At first they were assigned

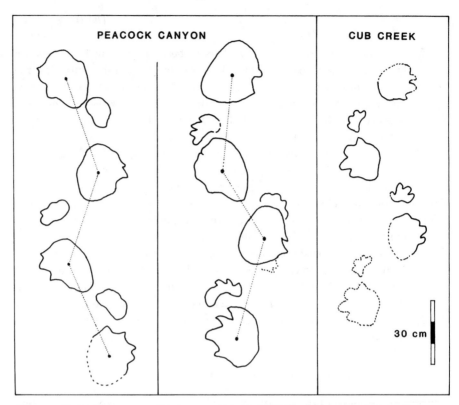

FIGURE 3.13 Large, possible prosauropod trackways from the Chinle Group of New Mexico (Peacock Canyon) and Utah (Cub Creek). These tracks may be assignable to *Tetrasauropus* (or to *Chirotherium*).

to the general *Chirotherium* track category, and they were inferred to be tracks of large crocodile-like aquatic phytosaurs. But we now favor an alternative interpretation (see the box, What Do Prosauropod Tracks Look Like?).

Similarly, *Apatopus* (meaning, deceptive foot) and *Coelurosaurichnus* (meaning, coelurosaur track) were both reported for the first time in the western United States at Peacock Canyon. *Coelurosaurichnus* was named on the basis of a Late Triassic track from Europe, and thought to differ from *Grallator*. We no longer believe this to be the case. In fact, ichnologists now think that *Coelurosaurichnus* is a synonym of *Grallator* (Fig. 3.14). *Coelurosaurichnus* cannot have been made by a coelurosaur because coelurosaurs, as now defined, did not appear until after the Triassic. During the time that Chinle sediments were being deposited there were many small dinosaurs (and even nondinosaurian archosaurs) that made *Grallator*-like tracks. The most probable candidate, and the one most frequently cited, is *Coelophysis*, or a close relative. The feet of this genus would match the tracks well. This case shows the potential danger of assigning a trackmaker name to a track when the trackmaker affinity is uncertain.

About eighty trackways of large to medium-sized animals have been found and recorded by our research group at the Peacock Canyon site, along with a huge number of lizard-like tracks referred to as *Rhynchosauroides* (figure 3.15). The name implies an affinity with a group of reptiles known from skeletal remains and named rhynchosaurs (beaked reptiles), but the implication is misleading. The name *Rhynchosauroides* originated when lizard-like traces were found early in this century in England, in deposits that also contained skeletal remains of rhynchosaurs. But rhynchosaurs are generally not lizard-like. In all probability *Rhynchosauroides* tracks were made instead by rhynchocephalians (meaning, beak-headed). Rhynchocephalians are lizard-like animals that include the modern Tuatara (genus *Sphenodon*), a "living fossil" that has held out against extinction on a few small islands along the coast of New Zealand. Tuatara tracks bear a close resemblance to the footprints of modern lizards, and indeed the animals are closely related. This conclusion that the tracks were made by a Tuatara-like animal is supported by the fact that *Sphenodon*-like and lizard-like remains both first appear in the fossil record during the Triassic.

If we are correct in correlating the abundant *Rhynchosauroides* tracks in Peacock Canyon not to rhynchosaurs but to rhynchocephalians, then the ancestors of today's Tuatara were very successful in western North America during the late Triassic. In places their tracks number several hundred per square meter. When combined with the footprint counts for other animals, thousands—even tens of thousands—of tracks are exposed at Peacock Canyon.

Such a large sample is useful as a census of the constituents of the animal community that inhabited this area in late Triassic times. The record suggests an abundance of aetosaurs, probable prosauropods (or phytosaurs, if the earlier interpretation holds), and lizard-like rhynchocephalians. But there were relatively few

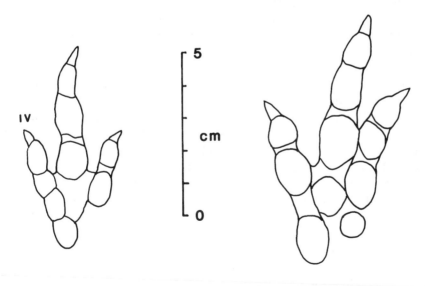

GRALLATOR   **"COELUROSAURICHNUS"**

FIGURE 3.14 *Grallator* and so-called *Coelurosaurichnus* from the late Triassic are actually the same track type according to Leonardi and Lockley (1995).

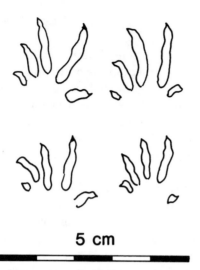

5 cm

FIGURE 3.15 *Rhynchosauroides* tracks, attributable to lizard-like trackmakers, are very abundant at some sites. *Left:* Drawing of *Rhynchosauroides* from the Chinle Group of New Mexico. *Right:* Photograph of mud-cracked surface with *Rhynchosauroides* and a single, large track of possibly a mammal-like reptile.

dinosaurs, mammal-like reptiles, and trilophosaurs. Geological evidence indicates that the tracks were all made along the banks of a river. In fact, the sinuous course of the river channel can still be discerned at the site (figure 3.16).

Peacock Canyon is not the only important late Triassic tracksite in this part of New Mexico. A recently discovered site in the area of Sloan Canyon about seventeen kilometers east of Peacock Canyon, has yielded a sample of about twenty-five trackways (figure 1.7), mainly of *Brachychirotherium* and a theropod dinosaur with tracks about 25 cm long. These are among the largest theropod tracks known from the late Triassic, and they indicate that theropods were becoming quite diverse by this time.

Tracks have also been found at Apache Canyon and Mesa Redonda in east-central New Mexico. The dominant track at Apache Canyon is *Brachychirotherium*. At Mesa Redonda this five-toed track type is less common, while three-toed tracks dominate. In the 1930s a number of track-bearing slabs from Mesa Redonda were sold to the University of Michigan Museum of Paleontology. One of the most interesting is still on display. It reveals unusual four-toed tracks—which we have recently identified as *Pseudotetrasauropus*, a track type first described from the late Triassic of southern Africa and attributed to a prosauropod. The recognition of this track type in the western United States (see figure 3.17) is significant because *Pseudotetrasauropus* was previously unknown in North America. The tracks are particularly interesting because they look like small versions of the large, four-toed, early Jurassic tracks *Otozoum*, which is generally attributed to a bipedal prosauropod. The *Pseudotetrasauropus* trackway from Mesa Redonda and another recently discovered near Dinosaur National Monument (figure 3.17) both reveal a small, bipedal animal. Such early evidence of bipedalism is noteworthy, and bipedalism is corroborated at the *Pseudotetrasauropus* site in southern Africa. The most obvious difference between

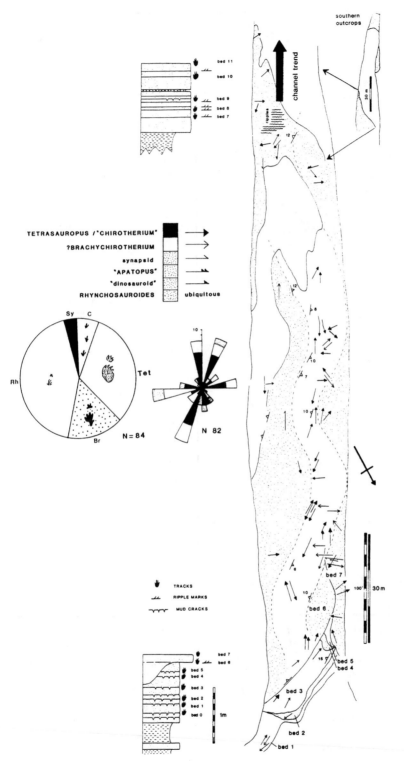

FIGURE 3.16 Peacock Canyon site map, with animal census based on track proportions. The sinuous river channel can still be recognized. The site also reveals at least eleven different beds of tracks in a thin stratigraphic interval.

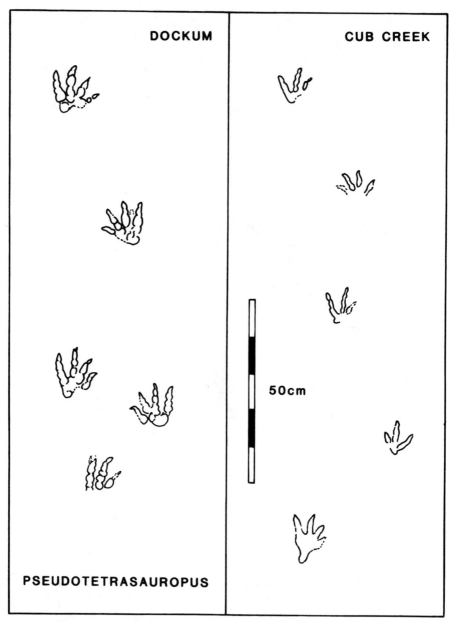

DOCKUM

CUB CREEK

50cm

PSEUDOTETRASAUROPUS

FIGURE 3.17 *Pseudotetrasauropus* tracks. The Dockum trackway was discovered in eastern New Mexico by a local rancher in the 1930s and identified by ourselves in 1991. We later discovered similar tracks at Cub Creek, near Dinosaur National Monument in northeastern Utah. These two specimens are the first of this type ever reported from North America. *After Lockley et al. 1992.* Ongoing studies of *Pseudotetrasauropus*-like tracks from around the world suggest that many small varieties are the same as a track type named *Kalosauropus* from the late Triassic–early Jurassic of southern Africa (Ellenberger, 1972).

What Do Prosauropod Tracks Look Like?

Prosauropods (meaning, before sauropods) are among the largest and best known of late Triassic dinosaurs, even though they are virtually unknown from skeletal evidence in western North America during this epoch. Their skeletal remains have primarily been found in Europe, southern Africa, and South America. Southern Africa is also rich in presumed prosauropod tracks; there the footprints are large and quite abundant. Paul Ellenberger, who published extensively on late Triassic footprints from southern Africa, named two similar and large track types as *Tetrasauropus* (four-footed reptile tracks) and *Pseudotetrasauropus*. He attributed both to prosauropods. Ellenberger found that *Tetrasauropus* tracks were generally large (hind feet of 40–60 cm long) and that they were made by quadrupedal animals, whereas *Pseudotetrasauropus* tracks were more variable in size (some with hind feet less than 30 cm long) and in most cases were made by bipedal animals.

Until recently nobody had recognized similar tracks in North America. However, at several sites in the Chinle Group we have discovered some small tracks that fit the *Pseudotetrasauropus* description (figure 3.17). We also conclude that *Tetrasauropus* is quite similar to the tracks referred to as *Chirotherium* sp. from Peacock Canyon (and previously attributed to phytosaurs). In fact, in 1987 when our research group first described the new tracks we discovered at Peacock Canyon (figure 3.13), we noted that "the trackways resemble those of small sauropods," thus causing us to "consider prosauropods as possible trackmakers." Clearly, the whole question of what prosauropod tracks look like requires further attention. Chapter 4 will revisit this issue.

the African and North American tracks is that the latter are much smaller, possibly indicating the presence of a hitherto unknown small prosauropod species in the western United States, or perhaps simply indicating smaller individuals of a species known from elsewhere.

A little later in the late Triassic, from the uppermost strata of the Chinle Group (known as the Sheep Pen Sandstone), we have found tracks in this same Eastern Region. In the vicinity of Kenton, Oklahoma, where the borders of Colorado, New Mexico, and Oklahoma come together, there are four small but important tracksites. One is in Oklahoma, another is in New Mexico, and two are in Colorado. Together these sites reveal a total of about a hundred tracks comprising the trackways of several dozen animals. All but four of these trackways are attributable to *Grallator*, indicating that the trackmakers were small, bipedal, carnivorous dinosaurs (figures 3.18 and 3.19). Two of the remaining trackways

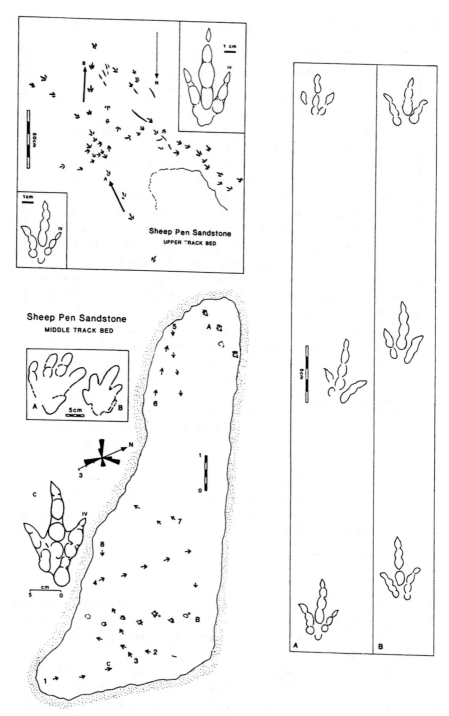

FIGURE 3.18 Small dinosaur tracks from the late Triassic Sheep Pen Sandstone, New Mexico. *After Conrad et al. 1987.*

FIGURE 3.19 Small *Grallator* tracks from the Sheep Pen Sandstone of Colorado (left) and New Mexico (right).

from the Oklahoma site were at first assigned to *Chirotherium*, but, in our view, they are probably equally well assigned to *Pseudotetrasauropus*. We base our view on the fact that *Pseudotetrasauropus* and another unknown track type occur at the *Grallator*-dominated tracksite in the same Sheep Pen Sandstone at the site in New Mexico.

Thus the eastern-most (and youngest) tracksites of the Eastern Region, in the Sheep Pen Sandstone, contain at least three track types, but no bones. The Kenton track record is therefore important in providing at least a hint of the dinosaur communities at the very end of the Triassic and how they differed from the animal communities recorded in earlier times, such as at the Peacock and Sloan Canyon sites of the Chinle Group. The increase in *Grallator* tracks evident in the Kenton area indicates a significant change in the makeup of reptile fauna at this time. Dinosaurs were relatively rare throughout most of the late Triassic, whereas a variety of other reptiles were common. Toward the end of this epoch—as a prelude to the Jurassic period—dinosaurs became increasingly abundant.

THE NORTHERN COLORADO PLATEAU REGION OF THE CHINLE

The third (and final) important group of late Triassic track-bearing localities is distributed mainly in northwestern Colorado, northeastern Utah, and southern Wyoming. We call this part of the Chinle the Northern Colorado Plateau Region even though parts of the area extend beyond the northern limits of the plateau.

Footprints were first reported from western Wyoming in the 1930s, when small dinosaur tracks were discovered and named *Agialopous wyomingensis* (meaning, track from the Popo Agie Formation of Wyoming; see figure 3.19).

In the 1960s a very important site was discovered just northeast of Dinosaur National Monument in northwestern Colorado. We revisited and studied the site in the 1980s, excavating a large collection of exquisitely preserved specimens that are now kept in the joint collections of the University of Colorado and the Museum of Western Colorado and in the U.S. Geological Survey collections. The tracks are important for three reasons. First, they help resolve problems associated with the interpretation of the enigmatic track *Agialopous*. Second, they provide a very large sample of the familiar track type *Rhynchosauroides*, including examples with skin impressions. Third, they provide us with the largest known sample of a third track type, *Gwyneddichnium* (named for a trace from the town of Gwynedd in Pennsylvania).

Agialopous tracks were made by a dinosaur about the size of a chicken or maybe a turkey. These tracks are similar to small *Grallator* tracks in general appearance, but differ in that a few examples display impressions of an apparent manus (figure 3.20). This is very unusual in the trackway of any theropod because theropods are traditionally regarded as bipedal animals. This *Agialopous* assemblage is thus exceedingly important. *Agialopous* tracks from this site in northwestern Colorado also include a few examples with metatarsal or heel impressions (figure 3.20), indicating where an animal squatted down on its haunches.

The original description of *Agialopous* provides an instructive case history in the dangers of initial interpretations. Edward Branson and Maurice Mehl made the original description in 1932. They mistakenly inferred that the difference between three-toed and four-toed *Agialopous* tracks of similar size was attributable to the former being front foot (manus) tracks and the latter being hind foot (pes) tracks. It is now abundantly clear that both the three- and four-toed tracks are the hind foot, or pes, tracks and that the concept of the *Agialopous* trackmaker is therefore ill-conceived since it refers in part to an imaginary conception of what the animal's front feet looked like. (In reality theropods, like most other dinosaurs, had small manus tracks in relation to their larger pes tracks). To add confusion to an already convoluted case, the type specimen of *Agialopous* was lost—making it impossible to compare it with subsequent track discoveries.

We suspect that *Agialopous* is probably the same thing as *Grallator*, a well known track type first named in the early nineteenth century by Edward Hitchcock to describe certain of the famous fossil footprints from the Connecticut Valley. Since that time *Grallator*, which clearly has historical priority over *Agialopous* as a scientific name, has been widely recognized in late Triassic and early Jurassic strata. Thus it is perhaps not too surprising to find abundant *Grallator* (or *Agialopous*) tracks in late Triassic strata in the Northern Colorado Plateau Region. We infer that such tracks are an indication of the proliferation of small dinosaurs in this region, and that they represent evidence of the same trend described in the Eastern Region.

Having just discussed the lizard-like track type *Rhynchosauroides* and its abundance in the Eastern Region, there is little to add here except that at some

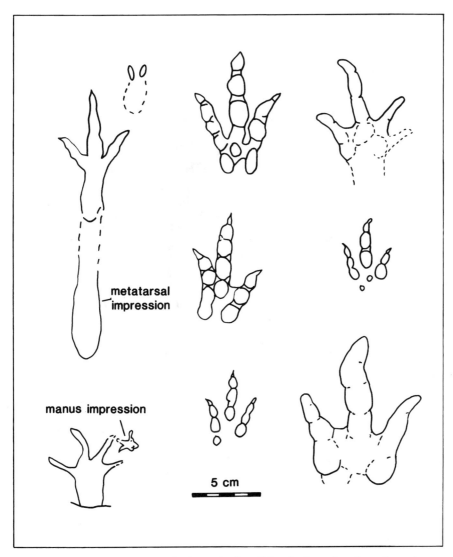

metatarsal impression

manus impression

5 cm

FIGURE 3.20 *Agialopous* tracks from the late Triassic deposits of northwestern Colorado. Note examples that have associated manus tracks and metatarsal impressions. *After Gillette and Lockley 1989.*

sites it is equally abundant in the northern Colorado Plateau region in rocks of the same (late Triassic) age. Without wishing to reveal undue prejudice toward *Rhynchosauroides* tracks or their trackmakers, we lovingly label them as the vermin of the Triassic. This characterization reflects their abundance and the fact that they often get in the way of a clear view of other tracks. Too many tracks can be too much of a good thing—especially when they obscure the track and trackway patterns of other rarer trackmakers.

Gwyneddichnium tracks (figure 3.21) were probably made by an aquatic or semi-aquatic reptile like *Tanytrachelos*, which is known from skeletal remains preserved in lake deposits from the eastern United States. (Remember, the type spec-

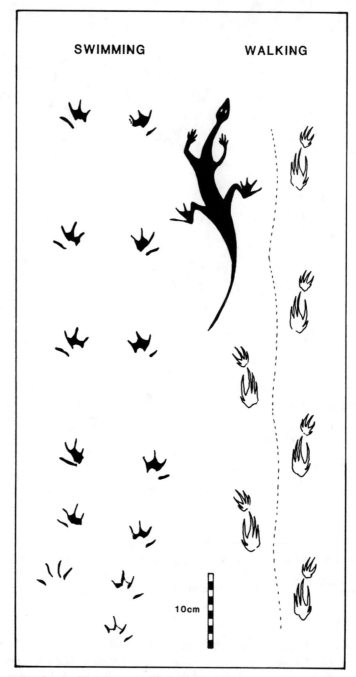

FIGURE 3.21 *Gwyneddichnium* tracks from the late Triassic of Utah (left) and Colorado (right) show swimming and walking progression, respectively. Note that the swim tracks reveal clear evidence of a web between three of the digits. *After Lockley et al. 1992).*

imen of the track *Gwyneddichnium* came from Pennsylvania.) The tracks provide the first record of such a trackmaker being abundant in western North America. In addition to these reptile tracks there are abundant traces of small inverte- brates, indicating possible insects and crustaceans living along the shores of a lake or waterway. All the vertebrate trackmakers were chicken-sized or lizard- sized, and some or all may have fed on the invertebrate trace makers.

In our own work, we have recently discovered many new tracksites in the late Triassic beds in and around Dinosaur National Monument. The discoveries represent the first substantial record of fossil footprints in this area so famous for fossil bones. These discoveries attest to the abundant vertebrate activity in the region in late Triassic times. Most of the track types known from the two other regions of the Chinle Group (the Southern Colorado Plateau Region and the Eastern Region) are now also known from the Dinosaur National Monument area. Not only does this show that diverse animal communities were widespread during the late Triassic, but it also shows the value of tracks in comparing or cor- relating different formations across the western United States.

Of the two dozen or more new sites found by us in the Dinosaur National Monument area, one of the most interesting revealed the trackway of a swim- ming *Gwyneddichnium* trackmaker. The tracks clearly show the animal's webbing between its toes. The traces apparently were made as it pushed against the bot- tom substrate as it paddled with synchronous strokes of its hind legs (figures 3.21 and 3.22). Not only is this a very rare example of tracks revealing the exact shape of fleshy tissue or soft parts, but it makes *Gwyneddichnium* one of the few track- ways that record both walking and swimming modes of behavior. This discovery supports the previous interpretation of *Gwyneddichnium* as having been made by an aquatic reptile like *Tanytrachelos*.

Another important Chinle site is the Cub Creek locality, very near the Dinosaur National Monument quarry in Utah. It has produced the trackways of more than fifty individual vertebrates from five stratigraphic levels. One of the maps we produced shows the trackways on the most extensively exposed sur- face (number 4, counting upward from the bottom). On this surface we found an association of *Grallator*, *Brachychirotherium*, and *Pseudotetrasauropus* (figure 3.23). At other levels we found trackways of *Pseudotetrasauropus* (figure 3.17) and the same trackway type, assigned to *Tetrasauropus* (*Chirotherium* sp.), that we first found at Peacock Canyon in New Mexico (figure 3.13).

Both *Pseudotetrasauropus* and *Tetrasauropus* are rare in North America, so the Cub Creek site represents a significant addition to the track record. As might be expected this type of large site adds much to the overall picture; we counted as many trackways at this one site as are found in the aggregate of twenty-five other sites in the vicinity. To date, our total census for Chinle sites in the immediate Dinosaur National Monument area is 111 trackways from 35 sites. (Most were discovered by our research group.) As shown in figure 3.24, all the vertebrate trackways fall into one of eight distinct categories and thus provide insight into the makeup of late Triassic animal communities in this area. Especially since vir- tually no tetrapod skeletal remains from the late Triassic have yet been found in this area, a track census is very valuable. (Remember that almost all the bones

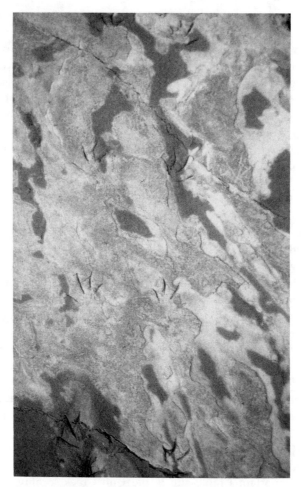

FIGURE 3.22 *Gwyneddichnium* tracks indicative of swimming behavior. (Compare with Figure 3.21).

from Dinosaur National Monument are Jurassic.) As shown in the next section, censuses can, in fact, be compiled for all known late Triassic tracksites in the western United States.

Interpreting Chinle Tracks and Track Habitats

The various Chinle Group deposits have a remarkable and varied track record. Altogether there are about a dozen distinct types of vertebrate tracks, apparently all attributable to reptiles. In some cases, notably southern Colorado Plateau sites, the tracks suggest animal communities that do reflect the known skeletal remains. In the case of the data from the Dinosaur National Monument area, the tracks tell a somewhat different story and add substantially to the very poor record of late Triassic fossil bones.

As illustrated in figure 3.25, the tracks can be used to provide a census of animal communities in the Colorado Plateau region during late Triassic times.

Overall, the picture is one of diverse reptile communities with few dinosaurs except in the very youngest deposits when the entire planet witnessed a remarkable increase in the number and diversity of dinosaurs. The wet monsoonal climate in what was then a geologic basin throughout the extent of what is today the Chinle Group evidently favored good trackmaking and track-preserving conditions. The geography and climate favored the success of aquatic reptiles (in lakes and rivers) as well as terrestrial reptiles. All in all, the late Triassic track record of the western states is one of the best in the world.

The Chinle Group spans the time from 208–225 million years ago, during the late Triassic. There are some gaps in the aggregate sedimentary record, marked as unconformities in figure 3.26. It is important to note that almost all the known tracks are in the uppermost unit, the Rock Point Sequence, which dates from about 208 to perhaps 215 million years ago. This might be explained by an abundance of wet substrates produced by monsoonal climates at this time. However, there is not much convincing evidence to suggest that latest Chinle time was that much wetter than early Chinle time. A better explanation pertains to how sediments accumulate.

As sedimentary basins form and fill, they undergo recognizable stages of basin evolution. Early on, when topographic differences between the newly formed basin and the surrounding uplands are most pronounced, the basins may begin filling, rather quickly, with gravels and coarse sands. This is because stream gradients are steep, causing rapid and high energy accumulation of sediment. Such dynamics create conditions that are not conducive to prolonged episodes of trackmaking and subsequent track preservation. As the basin matures, stream gradients lessen, and sediments become finer and accumulate more slowly. At this stage the water table rises toward the surface, creating lakes and a widespread distribution of environments suitable for track formation and preservation.

Probably the best supporting evidence for this geomorphic explanation for the concentration of Chinle tracks in the uppermost layers is the fact that similar track-rich facies have been observed in the upper parts of other terrestrial sedimentary successions where aggradation was also taking place. Thus it can be seen that, in addition to biological factors, sedimentological as well as climatic factors control track distribution.

Mass Extinctions at the Triassic-Jurassic Boundary

The transition from Triassic to Jurassic was marked by pronounced faunal changes on a global scale. The diverse reptile communities of the late Triassic disappeared and were replaced by communities dominated by dinosaurs. In terms of geological time this transition occurred fairly rapidly. There has been much debate about the factors that caused the end-of-Triassic extinctions. Some have postulated profound environmental changes as the cause; others envision dinosaurs out-competing their less advanced reptilian relatives for available food. Still others suggest extraterrestrial causes such as a devastating meteor or comet impact. At least one large meteor impact site, the Manicougan Crater, is known from eastern North America (specifically, the Hudson Bay region of

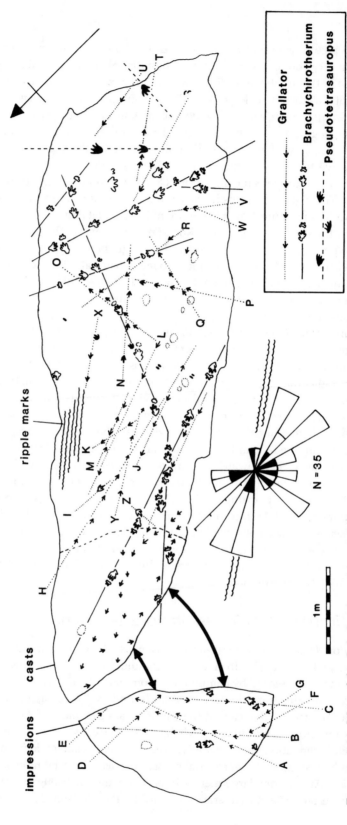

FIGURE 3.23 Map of the Cub Creek tracksite (late Triassic), Dinosaur National Monument area. The trackway orientation inset shows that a total of thirty-five trackway orientations have been recorded for *Grallator* (white) and other trackmakers (black). *After Lockley et al. 1992.*

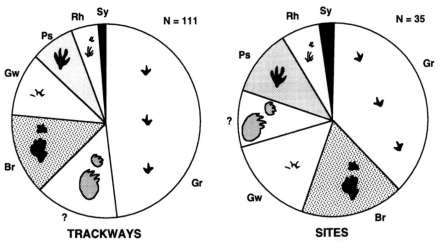

FIGURE 3.24 Census of late Triassic tracks from Dinosaur National Monument. *Top:* census based on number of trackways (= number of individuals). *Bottom:* census based on number of sites with particular trackway types. Gr = *Grallator.* Br = *Brachychirotherium.* Gw = *Gwyneddichnium.* Ps = *Pseudotetrasauropus.* Rh = *Rhynchosauroides.* Sy = synapsid. ? = ?*Tetrasauropus.* (Compare with Figure 1.7).

Canada), but the crater has not been conclusively shown to be exactly the right age to coincide with the Triassic-Jurassic boundary. In fact, the most recent evidence suggests that the meteor hit well before the global extinctions by which the end of the Triassic is defined.

Leaving aside the possible causes for such a turnover in global faunas, we will focus on the distinct changes that do appear in the vertebrate track record of the Triassic-Jurassic transition. A plot of different track types across the boundary reveals a distinct change (figure 3.25); few trackmakers of the late Triassic survived into early Jurassic time. In this respect, it is interesting to consider the difference between the high diversity vertebrate communities of Chinle time and the lower diversity communities whose tracks appear in the overlying Jurassic rock. (Chapter 4 will examine the details of this diversity change.)

In general, paleoecologists recognize that the replacement of a diverse fauna by one with very few species is a classic sign of a stressed ecosystem that has lost an ability to support much biodiversity. In such situations only a very few opportunistic species thrive and become dominant. So while diversity is low, a few species occur in abundance. This seems to fit the patterns observed at or near the Triassic-Jurassic boundary, when most familiar late Triassic trackmakers disappeared.

In fact it is only the ubiquitous track type *Grallator* that is found both in late Triassic and early Jurassic rocks of the western United States (figure 3.27). Even here some distinctions can be made as the Triassic *Grallator* tracks are generally smaller, at least in most cases, than those in the Jurassic. In the eastern United States different ichnospecies names have been assigned to these similar but different-sized tracks, but in the west we have not yet drawn such fine distinctions. We note only that very large three-toed theropod tracks, similar to *Grallator,* but

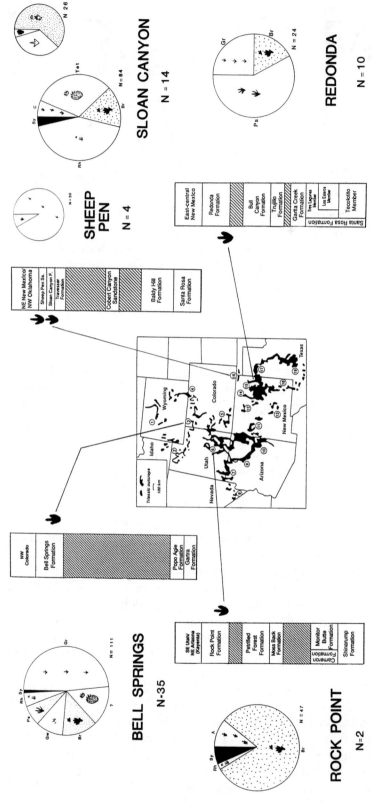

FIGURE 3.25 Censuses of late Triassic animal communities from the Upper Chinle Group based on tracks from many areas in the greater Colorado Plateau region, including the Shay Canyon, Peacock Canyon, and Dinosaur National Monument areas. Map and stratigraphy *after Lucas 1991.* Abbreviations as used in figure 3.24.

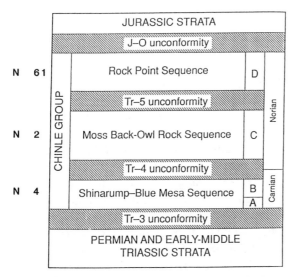

FIGURE 3.26 The distribution of tracksites in the Chinle Group is very unequal. Most tracksites occur in the uppermost sedimentary units: the Rock Point sequence. *After Lucas 1993.*

named *Eubrontes* (meaning true thunder) do not appear until at or near the Triassic-Jurassic boundary. We think of these larger trackmakers as heralding a new chapter in Mesozoic history.

Toward the end of the Triassic, in youngest Chinle Group sediments, small-sized *Grallator* tracks appear in association with a diverse assemblage of other archosaur tracks. Then in some of the very youngest Chinle layers (for example, the Sheep Pen Sandstone) these *Grallator* tracks become the dominant track type to the virtual exclusion of all other footprint types. We get the impression that the small dinosaurs were taking over. This impression is confirmed by the track record in the overlying strata of the Wingate Formation, where we find more medium-sized *Grallator* tracks, and very few other track types.

Although primarily the subject of chapter 4, the Wingate Formation has been considered Triassic by some geologists and Jurassic by others. The problem is that it represents a desert sand dune deposit without any fossils other than tracks. Most geologists agree that the Chinle Group below is Triassic and that the deposits above are Jurassic. Thus the boundary between these two periods can in theory be drawn anywhere between the base and the top of the Wingate Formation. For those who are skeptical that an impact event caused the extinctions at the Triassic-Jurassic boundary, the Wingate Formation provides evidence for a marked change in climate.

In the western United States there is widespread evidence of major environmental changes at the end of the Triassic epoch. The humid deposits of the Chinle Group and sediments from this so-called megamonsoonal epoch are overlain by desert sand dune deposits of the Wingate Formation and its equivalents. It is at about the time of this environmental change that *Grallator* tracks first appear in large numbers and begin to dominate track assemblages to the point of exclusion of most other track types. The Wingate also contains tracks of mammal-like reptiles. Although one could argue that the change in animal communities at this time was controlled by the abrupt transition from wet to dry climatic conditions, this hypothesis would not explain the fact that *Grallator* and *Eubrontes* tracks also appear in comparable abundance in many other parts of the world where there is much less evidence of sudden environmental change. This suggests that the abrupt climatic changes recorded in the western United States are not universal—nor are they the sole cause of the pronounced faunal changes that occur at the Triassic-Jurassic boundary in the western United States. In other words, the faunal changes at the Triassic-Jurassic boundary are real, and recognizable on a global scale. Except locally, they are evidently not an artifact of environmental changes or differences in the quality of the preserved track record.

Thus the increasing abundance across the boundary of the well-known theropod track *Grallator* and the appearance of *Eubrontes* marked a crucial phase in dinosaur history. Notably, the main radiation or diversification of theropods, and the evolution of large species in particular, did not get underway until the very end of the late Triassic and the beginning of the Jurassic (figure 3.28).

Although Aristotle advised us to study "all sorts of animals," he did not say that it would be easy, nor did he mention anything about the challenges involved in tracking Triassic archosaurs. Despite the bewildering diversity of Triassic tracks and the difficulty of confidently attributing them to trackmakers, at least one important point is clear. The track record clearly reflects a major radiation of

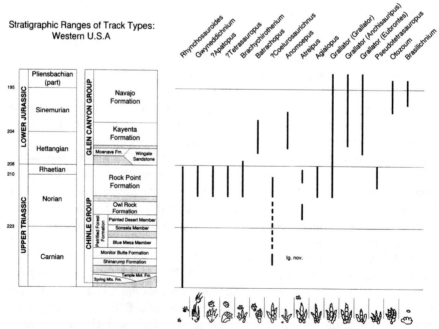

FIGURE 3.27 The change in footprint assemblages across the Triassic-Jurassic boundary clearly shows significant faunal turnover. Dinosaur tracks increase as other archosaur footprints decline or disappear.

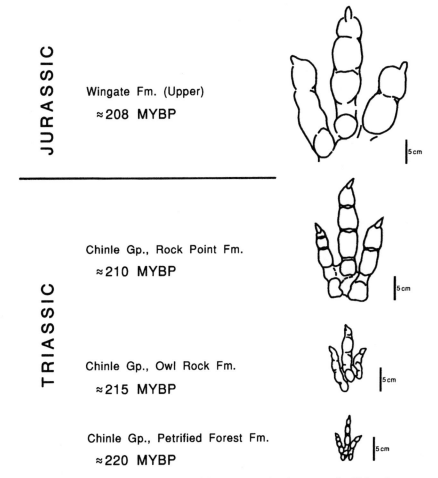

Wingate Fm. (Upper)
≈ 208 MYBP

5 cm

Chinle Gp., Rock Point Fm.
≈ 210 MYBP

5 cm

Chinle Gp., Owl Rock Fm.
≈ 215 MYBP

5 cm

Chinle Gp., Petrified Forest Fm.
≈ 220 MYBP

5 cm

FIGURE 3.28 Dinosaur tracks steadily increase in size near the Triassic-Jurassic boundary.

archosaurian reptiles during the Triassic and a widespread extinction of many archosaurs at the end of that period. The most successful group of archosaurs—the dinosaurs—weathered this crisis and went on to dominate the terrestrial realm in the subsequent Jurassic period. It is the story of these dinosaurian track-makers and their contemporaries that we turn to in chapter 4.

Exciting new discoveries made in the Triassic-Jurassic transition beds of western Colorado by Robert Gaston and Jennifer Schellenbach indicate the presence of the tracks of small mammals in both the Chinle and Wingate formations. These are very rare in the Mesozoic, owing to the small size and light weight of the trackmakers. Nevertheless, they are very important and will be published on in due course.

Diminutive dinosaur and spider in an early Jurassic desert.

4

DAYS OF DINOSAUR DOMINANCE I: THE JURASSIC

The fauna of that period, as shown by tracks alone, must have been unusually full, as we shall see when we come to describe the footmarks.

EDWARD HITCHCOCK, 1858

(IN REFERENCE TO THE JURASSIC)

■ The Jurassic, spanning 208–145 million years ago, is the middle period in the Mesozoic era. The Mesozoic as a whole is sometimes called "the age of dinosaurs." However, not until earliest Jurassic times did dinosaurs rise to prominence.

It is helpful to remember that dinosaurs were a diverse lot and that various groups arose at different times. The saurischians (or lizard-hipped dinosaurs) consisted of the theropods (the only carnivorous dinosaurs) and the herbivorous sauropods (sometimes called brontosaurs), along with their relatives the prosauropods. Theropods, and their narrow, three-toed trackways, are found throughout the age of dinosaurs. Prosauropods, on the other hand, are found only in late Triassic and early Jurassic deposits, so this is also where we find their tracks. Sauropods appeared in the early part of the Jurassic but did not become abundant until the late Jurassic. They were still around in the Cretaceous but in reduced numbers.

The other major group of dinosaurs, the ornithischians (or bird-hipped dinosaurs) were all herbivores. They consisted of the plated and armored dinosaurs (stegosaurs and ankylosaurs, respectively), the bipedal ornithopods (or bird-footed dinosaurs), and the horned dinosaurs or ceratopsians. Most of the ornithischians have a poor Jurassic track record; their tracks are much more abundant in the Cretaceous (chapter 5). We will, however, mention some of these groups in the Jurassic. We will also mention the tracks of other archosaurs, including the earthbound crocodiles and the flying pterosaurs. Other major groups of trackmakers that existed in the Mesozoic were mammal-like reptiles,

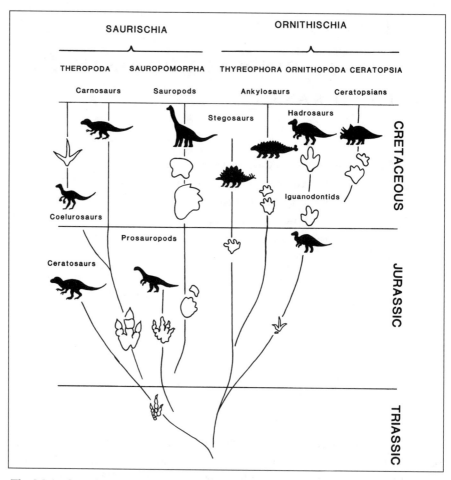

The Main Groups of Dinosaurs and Their Tracks.

birds, amphibians, and true mammals—though tracks of the latter two groups are virtually unknown.

The beginnings of dinosaur dominance are clearly evident from the Jurassic footprint record, which shows a remarkable abundance of dinosaurian tracks on a global scale. Both bone and track evidence suggest that the animal communities of the early Jurassic were dominated by carnivorous theropod dinosaurs, with a few ornithischian dinosaurs, prosauropods, and sauropod dinosaurs. Also present were various other reptiles of crocodilian and mammalian affinity.

One might ask how a vertebrate community could be dominated by carnivores. It would seem to run counter to the food pyramid of ecology, in which the bulk of the animal biomass is in the form of herbivores—at least in modern mammal communities (see chapter 7). In the case of the early Jurassic, perhaps the dominant (carnivorous) dinosaurs preyed mostly on each other, on other reptiles and vertebrates, on invertebrates, or on a combination of these groups.

The previous chapter already introduced some of the vertebrates that will be our focus here, but it is worth pointing out that Jurassic genera were anatomically quite different from their ancestors of the late Triassic. As we have seen, tracks of some of the bipedal carnivorous dinosaurs (theropods) began to get bigger at the close of the Triassic. Although sparse, the skeletal record of the early Jurassic confirms this trend by revealing such large creatures as *Dilophosaurus*, a form that is often cited as the probable maker of large *Eubrontes* tracks.

Increase in size was also one of the trends that affected other saurischians besides the theropods, namely the prosauropods and sauropods. Prosauropods are presumed ancestors of sauropods. Anatomically they differ from the sauropods in that they were smaller and in many cases were bipedal. Prosauropod tracks may be distinguished from sauropod tracks by their size also, and by their slender digit impressions. As we explained in chapter 3, the only probable trackway evidence (i.e., *Pseudotetrasauropus* and *Tetrasauropus*) for late Triassic prosauropods in the western United States indicates relatively small animals. By contrast, the appearance of *Otozoum* trackmakers in the early Jurassic suggests the existence of somewhat larger species of prosauropod, which we will describe shortly. No distinctly sauropod tracks are known from the western states before the late Jurassic. Footprints and bones of sauropods have been reported, however, from middle and even early Jurassic rocks at a few sites in other parts of the world, thus providing evidence of the evolution of true sauropods by this time.

The trend of increasing size was not universal by any means, even among the saurischians. Among other groups, such as the ornithischian dinosaurs, crocodilians, and mammals (which are believed to have arisen by the late Triassic), is evidence of various small- and medium-sized species. Early Jurassic ornithopod tracks known as *Anomoepus* indicate a turkey-sized animal. Similarly, tracks of mammal-like reptiles, probably tritylodonts, indicate animals no larger than medium-sized dogs. In fact, mammal-like reptiles were very much in decline by Jurassic times, and all known species and track types suggest groups that were smaller than many of their Triassic ancestors. True crocodilians are also an interesting group, first emerging in the Jurassic. Unlike modern species that are predominantly aquatic and large, many Jurassic species were generally smaller and better adapted to a terrestrial mode of life. We also encounter the tracks of the flying pterosaurs and can learn something of their size range and how they progressed when they alighted on land.

It is against this backdrop that we interpret the relatively diverse track record of the early Jurassic before moving on to examine the more obscure evidence from the middle Jurassic and finally on to the celebrated late Jurassic— "the age of brontosaurs." As we will show, in the western United States the early Jurassic track record is in many ways the most interesting and diverse of the entire Jurassic, as well as being one of the best in the world. The middle Jurassic paleontological record of vertebrates—especially skeletal evidence—is rather sparse worldwide, and the western United States is no exception. This epoch was a mysterious paleontological dark age. Though saurischian dinosaurs (theropods and sauropods) continued as successful groups at this time, the latter left no track or bone record anywhere in the west. By contrast, late Jurassic animal communities are well known, particularly for the abundance of sauropod skeletal remains. (Dinosaur National Monument is among the most famous of the late Jurassic sites.) The late Jurassic is sometimes even called "the golden age of brontosaurs." The track record of the late Jurassic is quite informative too, consisting of several dozen sauropod, theropod, and pterosaur footprint localities in the western states.

Considered overall, the early and middle Jurassic record of skeletal fossils is poor in the western United States, but the late Jurassic record is good. By contrast, the early Jurassic track record is now very good and the middle to late Jurassic track record is quite good but not outstanding. As we shall see, Jurassic tracks offer insights into the makeup of animal communities in this middle period of the Mesozoic. In many parts of the western United States, tracks provide a better record than can be gleaned from skeletal remains (figure 4.1).

Sand Seas of the Early Jurassic

The western states offer a plenitude of early Jurassic sites. Just in the part of the Colorado Plateau region where we have focused our own efforts during the past ten years—from the Four Corners area, north to Dinosaur National Monument—we have been able to document dozens of new sites. Many additional sites have been reported by others from western parts of the plateau and elsewhere, but these remain to be studied in detail.

The main sedimentary rock formations from the early Jurassic are collectively known as the Glen Canyon Group. In order of oldest to youngest they are:

Wingate Formation (Colorado, Utah, and Arizona).
Moenave Formation (Arizona and Utah)
Kayenta Formation (Colorado, Utah, and Arizona)
Navajo Formation (Colorado, Utah, and Arizona)

The ages of these formations have been debated for years, in large part because of the lack of fossils that could be used to more accurately determine their age. At one time all four of these formations were regarded as entirely Triassic. More recently, and as a few skeletal remains have been discovered, the pendulum has swung the other way; many geologists have concluded that the entire sequence, or at least the major part of it, is early Jurassic in age. (The early Jurassic is also known as the Liassic epoch.)

Footprints have been important in this debate. As demonstrated by the German tracker Hartmut Haubold in the mid 1980s, most of the known tracks from the Glen Canyon Group are Liassic in age. He even devised a scheme (see figure 4.2) showing three successive track zones, from oldest to youngest:

Brachychirotherium-Atreipus zone (late Triassic)
Batrachopus-Grallator zone (latest Triassic, near boundary)
Anomoepus-Eubrontes zone (early Jurassic)

This very generalized scheme works quite well around the world, but it requires some refinement before being applied to the western United States. Our most recent work suggests that at least the lowest part of the Wingate Formation is late Triassic. This means that we cannot simply draw the Triassic-Jurassic boundary at the base of the Wingate. There is no convenient coincidence in this

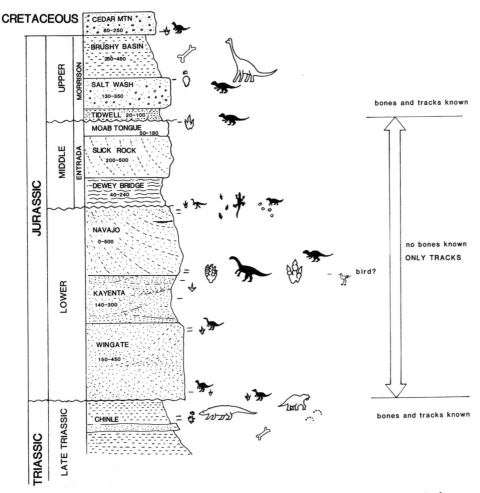

FIGURE 4.1 Jurassic track-bearing rocks of the Moab area represent a typical sequence in this central part of the Colorado Plateau region. Here, the vertebrates during a substantial chunk of Jurassic time are known only by their tracks. *After Lockley 1991.*

case of a major change in stratigraphy in the western states at exactly the time of the global faunal change by which geologists determined a break in geologic time periods.

The early Jurassic has offered some of the most famous fossil footprints of any type and any time. This is particularly true in the eastern United States, in the Connecticut Valley area, where Edward Hitchcock conducted his pioneer work in the early part of the nineteenth century. These "classic Liassic" studies led to the naming of the first dinosaur tracks ever documented in the scientific literature. Among the best-known examples are the ubiquitous *Grallator* and *Eubrontes* tracks attributed to theropod dinosaurs. These track types have been found at many sites formed in the early Jurassic and throughout the world.

The Glen Canyon Group of sedimentary rocks provides a classic example of desert deposits. Most obvious are the windblown, cross-bedded eolian sand-

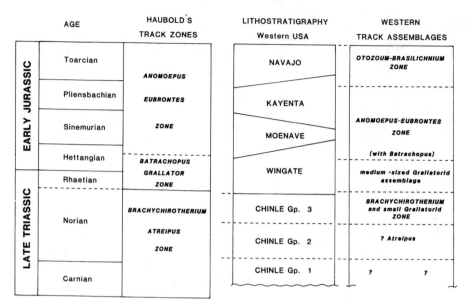

AGE		HAUBOLD'S TRACK ZONES	LITHOSTRATIGRAPHY Western USA	WESTERN TRACK ASSEMBLAGES
EARLY JURASSIC	Toarcian	ANOMOEPUS EUBRONTES ZONE	NAVAJO	OTOZOUM-BRASILICHNIUM ZONE
	Pliensbachian		KAYENTA	
	Sinemurian		MOENAVE	ANOMOEPUS-EUBRONTES ZONE
	Hettangian	BATRACHOPUS GRALLATOR ZONE	WINGATE	(with Batrachopus) medium-sized Grallatorid assemblage
LATE TRIASSIC	Rhaetian			
	Norian	BRACHYCHIROTHERIUM ATREIPUS ZONE	CHINLE Gp. 3	BRACHYCHIROTHERIUM and small Grallatorid ZONE
			CHINLE Gp. 2	? Atreipus
	Carnian		CHINLE Gp. 1	? ?

FIGURE 4.2 Late Triassic through early Jurassic track zones proposed by the German ichnologist Hartmut Haubold (1984), with refinements proposed by Lockley (1993) and the authors.

stones representing various types of fossil sand dunes. Arches National Monument contains spectacular examples of these. The dune habitat was so vast that it created sand seas or "ergs." Intercalated with these dune field or erg sediments are water-lain, interdune deposits that include various temporary, or ephemeral, lake deposits and sediments deposited by intermittent streams. The lake sediments are often referred to as playa or playa-lake deposits, and the stream sediments mainly represent sporadic runoff accumulations from flash floods. More often than not the stream channels (and the playa) would have been dry. Tracks are surprisingly common in many of the water-lain deposits—particularly those representing playa oases or watering holes that provided drinkable water and a source of moisture to sustain vegetation.

Desert Trackmakers of the Wingate Formation

The Wingate Formation is the oldest of the predominantly eolian deposits. It consists of massive, cliff-forming sandstones several hundred feet thick that rest atop the finer-grained siltstones and mudstones of the late Triassic Chinle Group. Until recently it was incorrectly thought, or assumed, that the Wingate Formation was virtually devoid of evidence of prehistoric life. However, tracks have been found at several dozen localities in Colorado, Utah, and Arizona.

Most of the tracks are found either very near the base of the formation or near the top. This may be a real phenomenon relating to the concentration of tracks in these zones, but it may be a biasing phenomenon related to the fact that the base and top of the formation are much more accessible than are the steep cliffs—and perhaps more easily eroded to allow investigation. A large number of Wingate tracksites are actually the result of huge rockfalls. The tracks themselves are found on the surfaces of sometimes house-sized boulders that have fallen from the sheer cliffs. In some cases we have been able to locate where these blocks fell from and thus determine the exact track-bearing level.

Wingate tracks are predominantly those of small- to medium-sized theropod dinosaurs of the type usually referred to as *Grallator*. These *Grallator* tracks are typically 15–20 cm long (figure 4.3), or about twice the size of the diminutive, yet abundant, *Grallator* tracks found in the upper part of the Chinle Group. But some basal Wingate localities reveal trackways of a large, wide-footed, wide-bodied animal of unknown affinity (figure 4.4). Perhaps they are some kind of prosauropod track—but this suggestion is tentative. They appear to be about the same size as the hind feet of the *Tetrasauropus* trackmaker discussed in chapter 3.

At Cactus Park near Grand Junction, Colorado, tracks near the top of the Wingate Formation include the large, three-toed variety known as *Eubrontes* (figure 4.5). These tracks were evidently made by a theropod that was much larger

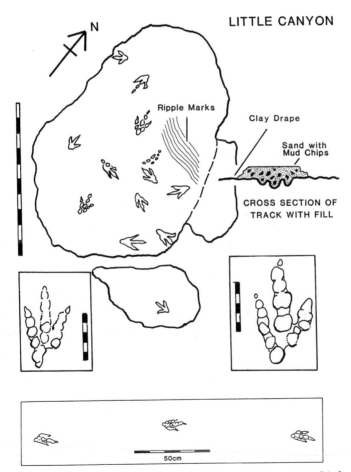

FIGURE 4.3 Typical *Grallator* tracks from the Wingate Formation, Little Canyon site (top) and Horsethief Bottom site (bottom). Both sites are near Moab, Utah.

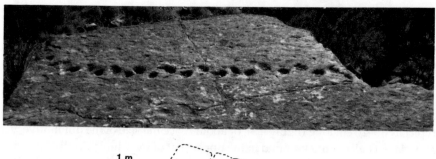

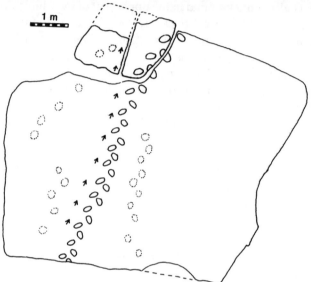

FIGURE 4.4 Rare trackway of a wide-bodied animal from the Wingate Formation (Dolores River Valley, western Colorado), with photograph of same.

than the *Grallator* trackmaker. Based on formulae that suggest that theropod foot length was about one-fifth to one-sixth of hip height, the *Grallator* trackmaker probably stood about 3 to 4 feet (about 1 meter) at the hip whereas the *Eubrontes* trackmaker was about twice this size. An animal that stood 6 to 7 feet (2 meters) at the hip probably had a head-to-tail length of 13 to 16 feet (4 to 5 meters).

Our census of about two dozen Wingate Formation tracksites reveals that *Grallator* is by far the most common footprint type, with occasional *Eubrontes* and rare tracks of the enigmatic wide-footed form shown in figure 4.4. We seriously doubt that all the *Grallator* tracks represent small individuals that grew up to be the makers of *Eubrontes* tracks. If this were the case, one would expect that *Eubrontes*-sized tracks would be much more common than they are. Rather, we suggest that the small trackmakers were representatives of a relatively small species that inhabited arid, desert environments. We also note that in contrast to the underlying Chinle Group, in which dinosaurs were few and other kinds of archosaurs were many, the Wingate evidence suggests a vertebrate community dominated by bipedal theropod dinosaurs. The pronounced earliest Jurassic dinosaur radiation is thus, in our view, showing up clearly in the track record.

In one or two areas, typical Chinle Group rocks are overlain by sandstones that generally lack fossils and so have been hard to identify or date accurately.

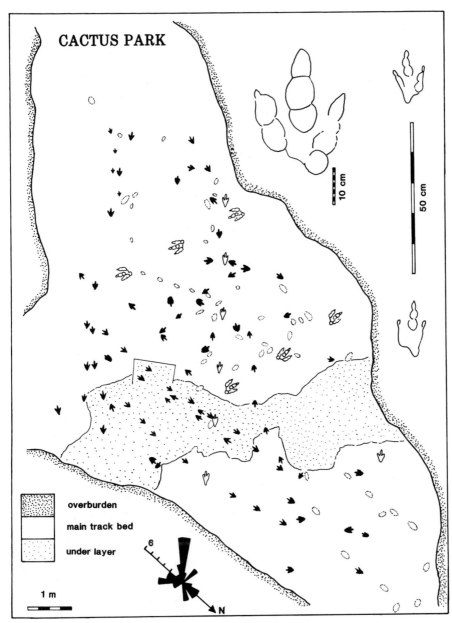

FIGURE 4.5 The early Jurassic tracksite at Cactus Park, Colorado, showing large *Eubrontes* and small *Grallator* tracks.   *After Lockley 1991.*

These sandstones may or may not be defined as equivalent to the Wingate. One example is the Sheep Pen Sandstone in the tri-state area of Colorado, Oklahoma, and New Mexico (see chapter 3 where it is included in the Chinle Group); another example is the lower part of the Glen Canyon Group at Dinosaur National Monument. In these two areas, we have nevertheless been able to find in the lower parts of the sandstones, just a few meters above typical Chinle Group rocks, *Pseudotetrasauropus* and *Brachychirotherium* tracks, respectively.

These tracks prompt us to date both deposits as very late in the late Triassic. In both cases tracks provide the only evidence of life in these rocks, and they offer new insights into the dinosaur communities that existed in these regions at that time. The Triassic-Jurassic boundary lies somewhere higher in the formations, but the exact place cannot yet be identified.

More Desert Trackmakers of the Moenave Formation

The Moenave Formation is confined to southern parts of the Colorado Plateau region—mainly northern Arizona, southern Utah, and possibly southeastern Nevada. The formation consists largely of sandstones, deposited both by wind and water. The lower if not greater part of the Moenave overlaps in time with the upper part of the Wingate. If one were to follow lower Moenave beds northward from southern Utah, for example, the water-lain sediments would gradually give way to thicker and more numerous dune deposits, finally passing laterally into the upper levels of the Wingate Formation.

Moenave tracks are known from Utah and Arizona and are predominantly attributable to *Grallator* and *Eubrontes*, as in the Wingate Formation (see appendix). Another distinctive footprint variety known as *Batrachopus* has been attributed to a crocodile-like reptile. The Moenave Formation has also yielded the skeletal remains of a primitive crocodile, *Protosuchus*, from beds only a few meters above the tracks. It is therefore reasonable to assume that *Protosuchus* or a *Protosuchus*-like animal was responsible for making *Batrachopus* tracks. Recent work suggests some alternate possibilities, however, as we shall explain later.

Batrachopus tracks also appear to be characteristic of early Jurassic sediments; they were first found in the historically famous deposits of the eastern United States and were named and described by Edward Hitchcock. Classic Liassic track assemblages with *Batrachopus*, *Eubrontes*, and *Grallator* have been reported from Dinosaur State Park in Rocky Hill, Connecticut. Hartmut Haubold therefore regards this track assemblage as a distinct track zone that is recognizable all around the world.

In the 1930s Roland T. Bird of the American Museum of Natural History discovered and photographed an extensive dinosaur tracksite in the Moenave Formation near Cameron in northwestern Arizona. These photographs were published by Edward Colbert in his popular dinosaur books, but the exact location of the site was lost. The site was finally rediscovered a half century later by Scott Madsen, a paleontologist now working at Dinosaur National Monument, and can now be studied in detail. To date, however, only one short abstract and a nontechnical magazine article have been published about these tracks, but the site is gaining fame as a result of being featured in a number of books and television documentaries. We and other trackers look forward to seeing a full description of the site and learning exactly which track types are found there.

Results of recent studies by our research group indicate that the Moenave Formation also contains tracks of mammal-like reptiles. These are very similar in appearance to the tracks of *Laoporus* from Permian sand dune deposits, and the *Brasilichnium* tracks, also found in sand dune deposits from the Early Jurassic Navajo Sandstone and described later in this chapter (figure 4.7). The presence of mammal-like reptile tracks in these three eolian deposits, indicates the track-makers' habitat preference for desert settings, and is supporting evidence for the concept of distinct vertebrate ichnofacies.

These tracks were only identified in 1994 and we are yet undecided as to whether they are best labelled as *Laoporus*-like or *Brasilichnium*-like, though both track types are very similar. What is clear however is that they are quite common and they are distinct from *Batrachopus* in detail. Superficially however they could be confused with *Batrachopus*, as discussed below (figure 4.26). We suspect therefore that it would be helpful to reexamine *Batrachopus* and *Brasilichnium*/*Laoporus*-like tracks from the Moenave to see how they differ in morphological appearance, preservation and distribution. If, as we suspect, the mammal-like reptile tracks are much more common than the purported crocodilian tracks (*Batrachopus*), then the concept of the *Batrachopus* zone will have to be reconsidered.

Kayenta Tracks and the Problem of "Provincial Taxonomy"

Like the uppermost Wingate and the Moenave, the Kayenta Formation is dominated by the three-toed tracks of large and small theropods. The best known tracksites are found near Tuba City in northeastern Arizona, near Moab in eastern Utah, and north of Kanab in southwestern Utah. Two kinds of tracks at the Tuba City site were described by Sam Welles in the early 1970s and given the names *Kayentapus* (meaning, footprint from the Kayenta Formation) and *Dilophosauripus* (meaning, footprint of *Dilophosaurus*); see figure 4.7. The latter name was chosen because skeletal remains of a remarkable crested theropod dinosaur, *Dilophosaurus*, were found only a few hundred yards away from the tracks in strata only fifty feet above the main track-bearing level. We should stress, however, that there is a significant body of opinion among trackers that these two new track names are just synonyms of better-known names like *Grallator*, *Anchisauripus*, and *Eubrontes*.

To begin, we ourselves think that *Grallator* and *Eubrontes* track types are distinct from each other. Like certain other trackers, we have trouble recognizing *Anchisauripus* as a distinct type because its only distinguishing feature seems to be that it is intermediate in size between *Grallator* and *Eubrontes*. (See the box, Different Species? Or Different Ages?) We note that both *Dilophosauripus* and *Kayentapus* are relatively large and *Eubrontes*-like in general appearance. We have examined the original specimen on which the ichnogenus *Dilophosauripus* was based, and it is not well preserved. *Kayentapus*, by contrast, is better preserved; it distinctly shows relatively slender digit impressions, pad impressions in the footprint, and wide angles of digit divarication. According to Robert Weems, *Kayentapus* is indeed a distinct track type that can be distinguished from *Grallator* and *Eubrontes*.

The best-known Tuba City tracksite is easily accessible just off Highway 160, by the exit to the village of Moenave, a few miles west of Tuba City. A roadside sign indicates the location of the tracks, as the site has become a tourist attraction (as has another roadside site a just mile to the east). The discoverer, Sam Welles, also described two other sites in the vicinity of Tuba City, and there are many others from the Kayenta Formation (and the Moenave) in this part of Arizona. One

Different Species? Or Different Ages?

Distinguishing three-toed theropod tracks of different shapes and sizes is a difficult problem. Some trackers, including Paul Olsen of Columbia University, believe that *Eubrontes* is just a large or adult version of *Grallator*. Others contend that the two track names do in fact distinguish different types of trackmaker.

Biologists know that as an animal grows up, different parts of its body grow at different rates. Thus an adult foot is not always exactly an enlarged version of a baby foot; the shape may be different. Change in track dimensions as an animal matures can be plotted on a graph. A straight line depicts the case in which no relative change of length and width dimensions takes place. Curved lines indicate otherwise.

Changes in the shape of any body part during the growth of an individual are referred to as allometry or allometric growth; such changes are very common. Figure 4.6 is a redrawing of a chart by Paul Olsen, which suggests that *Grallator, Eubrontes*, and an intermediate form (*Anchisauripus*) are indeed all part of a continuous allometric growth series. The tracks all seem to grade into one another, with no clear break.

There are many other examples of apparent allometric growth reported from the fossil record, especially among invertebrate creatures, such as clams and snails, whose extant relatives are known to change shape as they grow. In the case of clams, we can also estimate the animal's age by counting growth rings on the shell and by comparing growth rates with those observed in similar modern clam species. It is not so easy to do this with dinosaurs. No one has published very much convincing data showing the growth rates of dinosaurs. Even though some preliminary ideas on the growth rates of Cretaceous duck-billed dinosaurs have been proposed (see chapter 5), there is no information available on the growth rates of Jurassic theropods. We just do not know if their feet changed shape (grew allometrically) as they matured from juveniles to adults.

All that can be said at present is that *Grallator* and *Eubrontes* tracks are different in a number of respects. Most *Grallator* tracks are notably elongate (much longer than wide) owing to the narrow divergence (divarication angle) of toe impressions, and they generally display slender digit impressions with well-preserved digital pad traces. By contrast, *Eubrontes* tracks are much larger, with thick digit impressions, generally indistinct pad traces, and wider digit divarication angles. In the western United States, to our knowledge, no tracksites reveal obvious intermediate tracks between *Grallator* and *Eubrontes*. A diagram of western tracks similar to that constructed by Olsen for eastern tracks would show a clearer discontinuity in the case of most sites and formations. Basically, what is termed *Anchisauripus* in the East is missing or at least has not been identified in the West. We therefore conclude that *Grallator* and *Eubrontes* probably do not belong to the same species.

In support of this conclusion, we note that if Olsen's model were correct, it would imply that sites like Cactus Park (figure 4.5) revealed

(continued)

evidence of large numbers of juveniles, no intermediate-sized forms, and a few adults. By contrast, several dinosaur trackers have commented that tracks of juvenile dinosaurs tend to be rare. This may be for a number of reasons, including the fact that small tracks are not so easily made and preserved, or because dinosaurs grew rapidly and were therefore only small for a short portion of their lifespan. In any event, it would be highly speculative to suggest that assemblages of somewhat differently shaped small, medium, and large theropod tracks represent juvenile, intermediate-aged, and adult dinosaurs. Although we concede that Olsen's model has some merit in principle, and that small and large tracks of the same type should not be assigned to different species without adequate justification, we contend that the distribution of small and large tracks in the late Triassic and early Jurassic sediments of the West argues against his hypothesis.

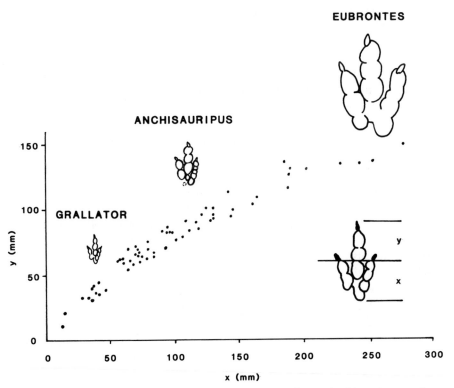

FIGURE 4.6 Paul Olsen's model proposing that *Grallator, Anchisauripus,* and *Eubrontes* form part of a growth series attributable to a single species. *After Olsen 1980.*

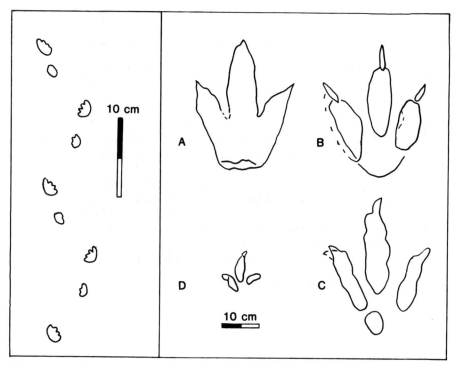

FIGURE 4.7 Tracks from the Moenave and Kayenta Formations, Tuba City area, Arizona. *Left:* trackway of a therasipid. *Right:* A and B: *Dilophosauripus*. C: *Hopiichnus* (= *Anomoepus*). D: *Kayentapus*. *After Welles 1971; Lockley 1986.*

of these sites has yielded small, three-toed tracks that Welles named *Hopiichnus* (in honor of the local Hopi tribe). This trackway is one of the most unusual ever described because of the exceptionally long step of 183—199 cm, relative to a foot length of only 10 cm. This suggests one of several possibilities: (1) the trackmaker had unusually long legs; (2) the trackmaker was running very fast, at 70–80 km per hour; (3) additional tracks in the sequence have been overlooked; (4) the trackway has been misidentified or misinterpreted in some way.

Welles suggested that the trackmaker was an ornithomimid, a lightweight, bird-like, gracile theropod that was certainly capable of running at a good speed. Ornithomimid skeletons are, however, first found in the Cretaceous. Several trackers have thus suggested that it is unlikely that early Jurassic trackmakers belonged to this group. We concur with the naysayers and offer the interpretation that *Hopiichnus* is simply an example of the better-known track *Anomoepus* that is relatively common in rocks of this age in many parts of the world. The general consensus is that *Anomoepus* represents an early ornithopod, not an ornithomimid. We have now found *Anomoepus* tracks at other sites in this region—the first report of such tracks in the western states—and so our interpretation of the Tuba City tracks is further supported. We have also examined replicas of the original tracks studied by Welles, and we conclude that most tracks in the trackway sequence are poorly preserved. This suggests that the trackway may be incomplete and the very long step measurements the result of missing tracks in the sequence.

This particular example of giving three new names to tracks from a localized area, when suitable names already exist, is what some authors have called "provincial taxonomy." It is a perennial problem in paleontology because it obscures similarities between tracks (and other fossil evidence) from different regions that should be recognized and used to advantage. In this case, recognition that the Kayenta tracks are the same as those from the classic Liassic assemblages of New England would be an additional demonstration that many tracks from both the western and eastern United States are essentially the same. And only in this manner will it be possible to conduct biostratigraphic correlation between the widely separated track zones of the East and West.

Kayenta Formation tracks are also known from the vicinity of Moab, Utah. Here too they consist predominantly of the tracks of three-toed carnivorous dinosaurs, notably, *Eubrontes*. One large slab is currently on display at the Dan O'Laurie Museum in Moab (figure 4.8).

Another recently discovered site occurs in the Lisbon oil field area at the boundary between the Kayenta Formation and the underlying Wingate Formation. The Moenave Formation is absent in the Moab area, so the boundary

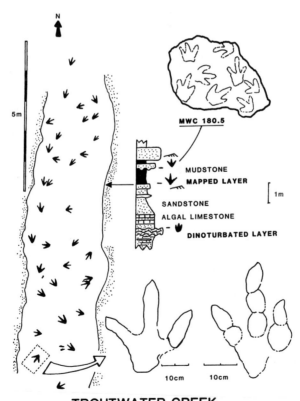

TROUTWATER CREEK

FIGURE 4.8 Tracks from the Kayenta Formation, Troutwater Creek, near Moab, Utah. Tracks resemble *Eubrontes* and *Kayentapus*.   *Drawn from a numbered slab on display at the Dan O'Laurie Museum in Moab.*

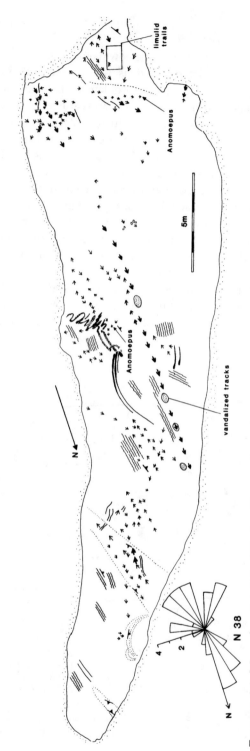

FIGURE 4.9 Map of tracks at the Wingate-Kayenta boundary, Lisbon oilfield site near Moab, Utah. Map shows details of location of *Anomoepus* trackways, ripple marks (wavy lines), and vandalized tracks.

is really an unconformity. At this site several hundred three-toed tracks have been recorded on a ripple-marked surface that appears to represent a lake shoreline (figure 4.9). In addition to typical *Grallator* tracks there are a number of three-toed and some four-toed footprints that show wide angles between the two outer toes and are therefore very much like bird tracks.

These bird-like tracks are of two types. A relatively small, four-toed impression can be assigned to *Anomoepus*, the distinctive classic Liassic track type first reported from the Connecticut Valley but now also known from Europe and Africa. These tracks usually reveal a pigeon-toed inward rotation of individual tracks (figure 4.10). However, most authorities attribute *Anomoepus* tracks to ornithopods, not to birds. This is because examples from the Connecticut Valley occasionally show five-toed front footprints, as well as heel (metatarsal) and pelvic impressions. This shows that the animal sometimes switched from bipedal to quadrupedal pro-

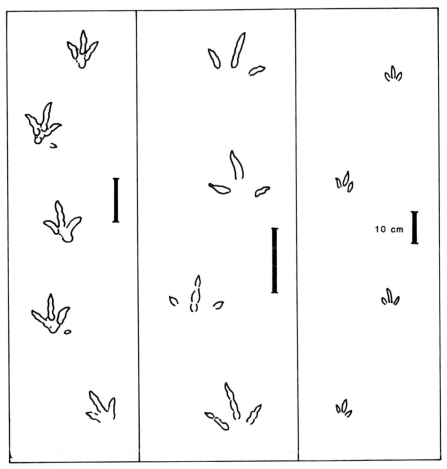

FIGURE 4.10 *Anomoepus* trackways from the Lisbon oilfield site. Note wide divarication of digit impressions. Note also, in the left trackway, two tracks that show marks made by digit I.

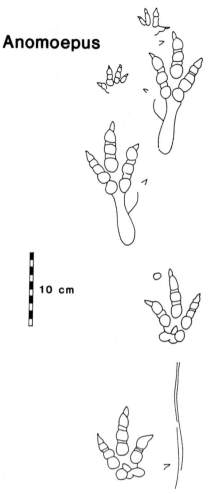

Anomoepus

10 cm

FIGURE 4.11 *Anomoepus* tracks from the eastern United States sometimes reveal evidence of quadrupedal locomotion and crouching. Specifically note the heel (metatarsal) and manus impressions in the third and fourth tracks. *After Lull 1953.*

gression and even to a stationary crouching position (figure 4.11). It is the occasional presence of five-digit impressions on the front foot that makes the ornithopod trackmaker interpretation most persuasive. In contrast, theropods from this epoch appear to have been habitually bipedal, and we know, based on skeletal evidence, that they usually had only three or occasionally four manus digits.

The second type of bird-like tracks found at the Lisbon site is characterized by extremely narrow digit impressions and a wide splay (divarication) of the digits—both features are characteristic of birds. This track type is very similar to a track described from the earliest Jurassic of Hungary and named *Komlosaurus* (after the town of Komlo), another described from China under the name *Schizograllator* (meaning, "split" *Grallator*), others named *Trisauropodiscus* and *Trisaurodactylus* from southern Africa, and yet another unnamed, relatively large and slender-toed track type known from the early Jurassic of Morocco (figure 4.12). When considered in relation to *Grallator* and *Eubrontes,* the Lisbon site tracks do appear to be distinct footprint types. This would be consistent with the observations of Robert Weems that not every footprint in the western states at this geologic time can be expected to fall under the *Grallator* or *Eubrontes* classification.

Although we may refer to a new type of print that appeared in the early Jurassic as being "bird-like," this does not mean that the trackmakers were the ancestors of birds. Conventional paleontological wisdom argues against the presence of early Jurassic birds. True birds, like the feathered genus *Archaeopteryx,* became established by late Jurassic times. Recently there have

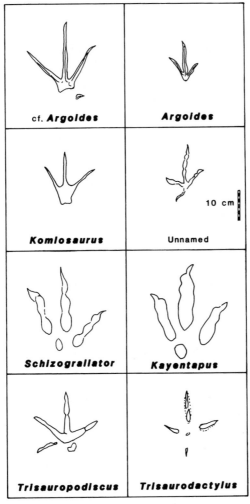

FIGURE 4.12 Slender-toed dinosaur tracks from the early Jurassic. Clockwise from top left: *Argoides* from Morocco and New England; unnamed track from eastern Utah; *Kayentapus* from northeastern Arizona; *Trisaurodactylus* and *Trisauropodiscus* from southern Africa; *Schizograllator* from China; and *Komlosaurus* from Hungary.

been controversial reports of at least one older bird or bird-like genus (*Protoavis*) from late Triassic strata. We suggest, therefore, that it is not impossible that the early Jurassic track record may contain tracks of birds or very bird-like dinosaurian species. As in the case of dinosaurs, the oldest evidence of a new lineage may be a track rather than a body fossil.

As an historical aside, we note that the three-toed tracks discovered and studied by Edward Hitchcock in the Jurassic strata of New England were first interpreted as the footprints of giant extinct birds. This made perfect sense at the time because dinosaurs had only recently been discovered, and they were then thought to have looked like giant lizards. Later, as dinosaurs became much better known and it was recognized that they did have distinct bird-like characteristics, it became clear that the bird-like Connecticut Valley footprints were the tracks of theropods and other dinosaurs. Most of Hitchcock's interpretations were questioned. We say "most" because Hitchcock was familiar with *Anomoepus* trackways that showed small front footprints. This led him to infer that the probable trackmaker had been a bird-like reptile. This was a prophetic conclusion at a time when the bird-like characteristics of dinosaurs had not yet been recognized.

In this respect it is very interesting to note that recent studies show the birds as being closely related to and descended from theropod dinosaurs, thus making Hitchcock's conclusions essentially correct. It may soon be time to look at this problem again and decide which tracks are really bird-like and which are not— and perhaps even which tracks represent the first true birds. Some of these problems may be solved in the future by comparing tracks with known body fossil remains. In recent years the Kayenta Formation of the Four Corners area has yielded the skeletal remains of mammals, turtles, crocodilians, and several different dinosaurs (theropods, prosauropods, and ornithischians).

For the tracks at Tuba City and at certain sites near Moab there is some doubt about where exactly in the stratigraphic succession they occur. More precisely, there is sometimes uncertainty or disagreement among geologists about where to draw the boundary between formations. As many track-bearing layers fall at or near boundaries, changing opinions about stratigraphy can cause a particular track bed to be reassigned to an altogether different formation. This is particularly true of the formations within the Glen Canyon Group because the strata look quite similar from site to site.

For example, when we first learned of the Lisbon site, we were told that it was "in" the Wingate Formation. This interpretation is understandable because the tracks are in fact impressed into the uppermost unit of the Wingate Sandstone. However, the Wingate is predominantly an eolian dune deposit, and the tracks occur in a layer that also contains ripple marks—evidence that water reworked or redistributed the Wingate sediments. The evidence thus suggests that the top of the Wingate was wetted and reworked at the beginning of Kayenta time—that is, after the end of Wingate deposition. It is also pertinent to note that the track assemblage is different from that normally found in the Wingate and thus probably indicative of a much younger footprint zone (see figure 4.2 for a diagram of stratigraphy).

Turning now to other instances of possible confusion, it has been suggested that some of the Kayenta Formation tracks near Tuba City are in fact in the

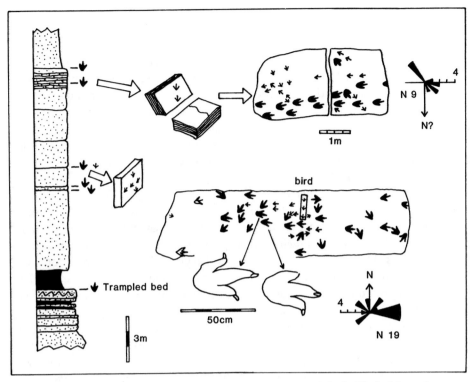

FIGURE 4.13 Tracks from the Navajo Formation near Moab, Utah. Note six track-bearing levels. Tracks from steep cliffs of the Navajo are best seen on blocks that have fallen onto Kayenta outcrops below. Tracks labeled 'bird' are those illustrated in figure 4.16.

underlying Moenave Formation strata. This simply reflects differences of opinion as to where the boundary should be drawn and does not change the physical location of the track beds. Similarly, near Moab some so-called Kayenta Formation tracksites are actually in the lower part of the overlying Navajo Formation. Perhaps the best example of misidentification of stratigraphy occurs at an accessible, posted roadside location seven miles along the Potash road (Highway 279), south of its junction with Highway 191. Tracks occur in large blocks that have fallen out of the Navajo Formation cliffs (figure 4.13). Several reports describe these as tracks in the Kayenta Formation because the blocks came to rest at a lower level on the hillside, among Kayenta rocks. Recent work in the vicinity of Canyonlands and Lake Powell by our group reveals many tracksites in the upper part of the Kayenta Formation, just below the overlying Navajo Formation.

An Abundance of Tracks in the Navajo Formation

In the western United States there can be little doubt that the most interesting and varied footprints of the early Jurassic occur in the Navajo Formation. Among the track types identified are the footprints of large and small three-toed dinosaurs, prosauropods, mammal-like reptiles, and various other vertebrates, along with some invertebrate traces.

The Navajo itself is, from a geological perspective, an important and controversial formation. Many sedimentologists regard it as the classic example of a desert sand sea deposit. At Arches National Park, for example, signs proclaim that large expanses of Navajo Sandstone represent fossil sand dunes. Although there is no doubt that the Navajo Sandstone does represent a fossil sand sea or "erg," a few geologists once argued that dunes and dune-like features could have accumulated under water. The majority of geologists argued convincingly against this theory for the Navajo, pointing out, among other things, that the deposit reveals many fossil footprints.

Recent work shows that the Navajo is indeed a classic desert sand dune deposit. Locally, however, can be found evidence for small oases or playa lakes represented by distinctive thin layers of grayish limestone that often contain remains of algae. These distinctive layers sometimes yield fossil wood and often reveal tracks. Although rare, the Navajo has also yielded skeletal remains of crocodilians (a small primitive *Protosuchus*-like crocodilian), mammal-like reptiles known as tritylodonts, and dinosaurs including the chicken-sized theropod *Segisaurus* and the small prosauropod *Ammosaurus*.

Many of the footprints known from the Navajo Formation come from flat-lying strata that represent playa lakes or various water-lain or interdune deposits. Others, however, are found on the surfaces of inclined strata known as cross sets or fore sets. These inclined layers represent the slip face or avalanche

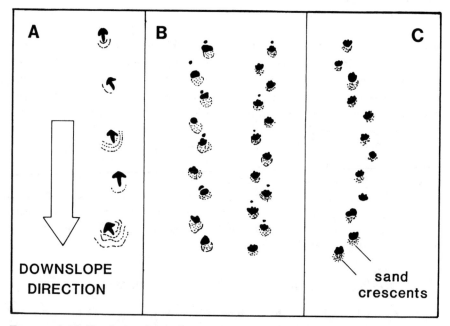

FIGURE 4.14 Tracks made on sloping dune faces. Note the sand crescents formed on the downslope side of each track. A and C: based on early Jurassic of the western United States. B: based on the Jurassic of Brazil. B and C: = *Brasilichnium*.

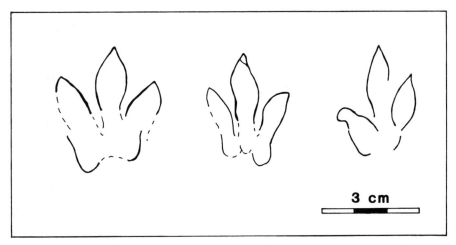

3 cm

Figure 4.15 Enigmatic small tracks collected by Jim Jensen from the Navajo Formation near Moab (CU-MWC 182.2).

face on the leeward side of advancing dunes. Tracks made on these sloping surfaces were quickly covered as the dunes advanced. These tracks also reveal characteristic bulges or sand crescents that would have formed on the downslope side (figure 4.14). Such bulges have, in fact, been used to determine the downslope direction in ancient dune deposits.

The Navajo Sandstone is exposed over a vast area of the western states. In Idaho, Wyoming, Colorado, and Utah parts of the Navajo Formation are referred to as the Nugget or Navajo-Nugget. In Arizona, Nevada, and California parts of it are referred to as the Aztec Sandstone. Tracks are found in many of these areas and have recently been studied, though only preliminary accounts have been published.

In the lower part of the Navajo Sandstone, many of the tracks are of three-toed carnivorous dinosaurs. These footprints show a wide size range, from less that 10 cm to almost 50 cm in length. The smallest tracks include two diminutive tracks collected by Jim Jensen along the Potash road south of Moab (figure 4.15). The significance of these tracks remains elusive because their exact point of origin is unknown. However, if they are indeed dinosaur tracks then they must represent a sparrow-to-blackbird-sized species or a baby or a hatchling. The same location contains a number of other small, three-toed tracks displaying slenderer and hence very bird-like digit impressions (figure 4.16). These have recently been named *Trisauropodiscus moabensis*, meaning, three-toed track from the Moab area and denoting that they are similar to *Trisauropodiscus* tracks (with different ichnospecies names) from early Mesozoic strata in southern Africa. When *Trisauropodiscus* was originally named by tracker Paul Ellenberger, he observed that the footprints were extremely bird-like and so he coined the name *T. aviforma* (meaning, bird-like in form) for some of the African examples. Because it is now believed that birds evolved from small theropod dinosaurs, it may be difficult, at least in theory, to distinguish very early bird tracks from those of their small dinosaurian peers and predecessors. We know from body fossils, however, that birds were present at least as early as the Jurassic period. Thus, we might expect to find their tracks appearing at that time.

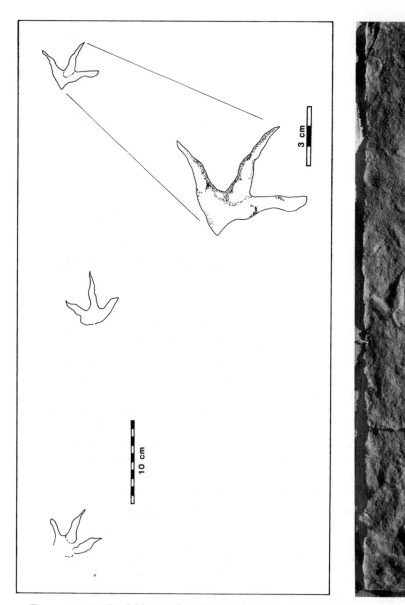

FIGURE 4.16 Bird-like tracks, *Trisauripodiscus moabensis*, from the Navajo Formation near Moab (after Lockley et al. 1992) with photograph of the rubber mold of same.

Recent studies by our research group reveal the presence of at least one Navajo tracksite that yields a medium-sized three-toed track that is quite distinct from *Grallator*, *Eubrontes*, and *Trisauropodiscus*. Our preliminary conclusion is that the trackway is similar to *Anomoepus* and that the trackmaker, therefore, may have been an ornithopod. Similar tracks also occur in younger Late Jurassic deposits of the Morrison Formation. This suggests that ornithopods, though rarely represented in the Jurassic track record, were present in Kayenta through Navajo to Morrison time. Again such evidence shows us that the Jurassic track record needs further exploration and study.

Another interesting and very different kind of fossil track from the Navajo Formation is a large, four-toed variety known as *Otozoum* (meaning, giant animal). The tracks in figures 4.17 and 4.18 were discovered near Moab. Although once thought to be relatively uncommon in the region, they are in fact quite common. This is perhaps no big surprise as *Otozoum* was one of the first track types described by Edward Hitchcock from early Jurassic strata in New England. Hitchcock was puzzled by these giant animal tracks because they usually indicated a bipedal animal. In fact, although there were reports of a small, five-toed front footprint in association with an *Otozoum* footprint, it was never convincingly demonstrated that the footprint was part of a regular trackway sequence made by an animal progressing on all fours. In fact, just recently this purported five-toed front footprint has been reinterpreted by Paul Olsen as two superimposed tracks of small, three-toed dinosaurs. Thus the overall evidence today is even stronger that the *Otozoum* trackmaker was always bipedal.

There has been considerable debate about the probable identity of the *Otozoum* trackmaker. Some have claimed it was a prosauropod dinosaur—others

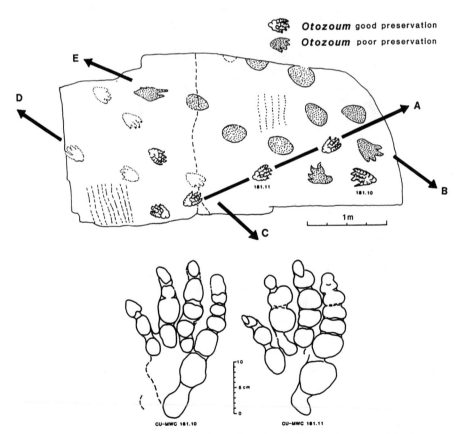

FIGURE 4.17 *Otozoum* trackways from the Navajo Formation, near Moab. Note that trackways show variable step lengths, and all reveal bipedal progression. *After Lockley 1990.*

FIGURE 4.18 Photograph of *Otozoum* tracks from near Moab. (See Figure 4.17 for details).

a crocodilian, or even a large ornithopod. Various crocodiles are known to have existed at this time, and varieties with terrestrial adaptations even frequented desert environments. But no body fossils have yet been found that are large enough to make *Otozoum* tracks. For this and other reasons we infer that the trackmaker was probably a prosauropod—a view that is becoming increasingly popular. These animals included representatives that were bipedal. Alternatively, the trackmaker may have stepped on its front footprints with its hind feet, thus appearing bipedal. Sauropods did this sometimes, depending on their speed, but crocodilians generally did not.

Recent discoveries of two new *Otozoum* tracksites in the Navajo Sandstone of the Lake Powell region, and other sites at Dinosaur National Monument, show that *Otozoum* tracks are quite common in the western United States. Moreover, at almost all the known localities, now numbering about twelve, *Otozoum* track-ways always appear to have been made by bipeds. The discoveries also suggest that the presumed prosauropod trackmaker responsible for *Otozoum* tracks pre-ferred relatively arid interdune environments. This inference is consistent with the hypothesis that sauropods were associated with dry, rather than humid environments. At Lake Powell, *Otozoum* trackways are found in interdune deposits and are mainly oriented perpendicular to the prevailing wind direction, which can be discerned from the structure of the dunes themselves (figure 4.19). The trackways relationships to the fossil dune deposits suggests that the trackmakers may have been walking parallel to the avalanche or slip face of the dune on the lee side. The *Otozoum* tracks at Lake Powell are also found in association with many very small theropod tracks. Elsewhere, Navajo tracks from interdune deposits include large *Eubrontes*-like footprints, which presumably could have been made by theropods that preyed on the *Otozoum* trackmakers. This suggests ancient animal communities that consisted of prosauropods, large theropods, and various much smaller bipeds (probably mainly theropods).

Whatever the origin of *Otozoum* tracks, we know of at least one other trackway type that has also been attributed to a prosauropod. This is a trackway of a four-footed animal referred to as *Navahopus* (meaning, footprint from the Navajo Formation). This trackway (figure 4.20) was made by a much smaller animal than the *Otozoum* trackmaker—possibly a small prosauropod. The *Navahopus* trackway was found in Arizona and has been interpreted as a prosauropod walking up the inclined slope of a sand dune. Donald Baird, who proposed the prosauro-

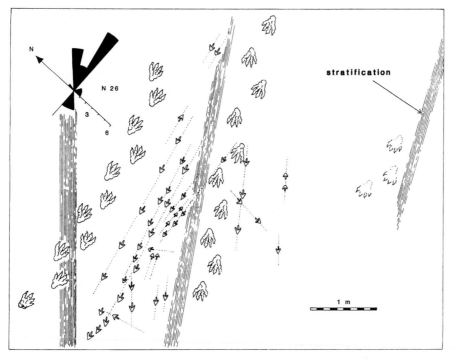

FIGURE 4.19 *Otozoum* and theropod trackways from the Navajo Formation at Lake Powell. The orientation (inset) indicates the trackmakers were walking parallel to the slipface of the dune.

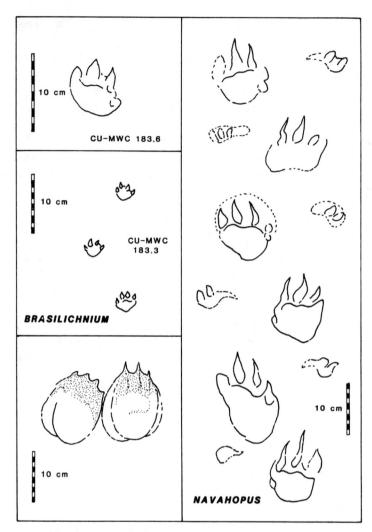

FIGURE 4.20 *Navahopus* tracks from the Navajo Formation, Arizona, based on drawing of the type specimen. Large *Brasilichnium* tracks are shown on the left, for comparison.

pod interpretation and coined the name *Navahopus*, noted that the front footprint appeared to show the impression of digit I—the thumb or "pollex"—which is a distinctive characteristic of prosauropod front feet.

We note here that all other prosauropod trackways (*Otozoum*) from the Navajo Sandstone are associated instead with originally flat-lying, water-lain, interdune deposits. This could suggest that the *Navahopus* trackmaker had different adaptions or habitat preferences from the larger *Otozoum* trackmaker. Another, simpler but perhaps more controversial, explanation is that the trackmaker was not a prosauropod at all, but instead was a relatively large, mammal-like reptile of the type responsible for making *Brasilichnium* trackways. In support of this suggestion we note, from an examination of the type specimen (figure 4.20), that the trackway and presumed pollex impressions are not particularly well preserved and that the foot size is not much larger than the largest known "mammaloid" (*Brasilichnium*) track specimen, discussed next. We also point out that most small trackways assigned to prosauropods indicate bipedal animals. We stress, however, that our alternative interpretation is tentative and that Baird's original interpretation still remains credible, especially with respect to the purported thumb (pollex) impression.

Is It Mammaloid?

In the upper part of the Navajo Formation, especially near Moab and to the northeast near Dewey Bridge in eastern Utah, abundant trackways of a small, quadrupedal, mammal-like creature have been found. The tracks (figures 4.21 to 4.23) have been described as footprints of a small, broad-footed trackmaker because the oval-shaped foot is wider than long. We recently studied the sites and have come to interpret these as the tracks of so-called mammal-like reptiles, probably tritylodonts. These footprints (figures 4.22 and 4.23) are very different from all known dinosaur tracks, being generally small and wide with at least four, and often five, recognizable digit impressions on the hind footprints. They have been found in South America and southern Africa too. In each of these locations the discoverers were largely unaware that virtually identical tracks existed on other continents. Again the problem of "provincial taxonomy" has arisen and trackers have given the tracks different names in different areas.

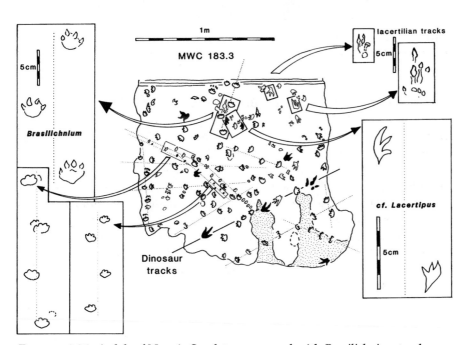

FIGURE 4.21 A slab of Navajo Sandstone covered with *Brasilichnium* tracks that is on display at the BLM office in Moab (CU-MWC 183.3). *After Lockley 1990, 1991.*

FIGURE 4.22 Photograph of *Brasilichnium* tracks from slab at BLM office in Moab. (See figures 4.20 and 4.21 for comparisons and context.)

The most recent and thorough published study was based on South American material. The tracks were named *Brasilichnium* (trace from Brazil). They are a broad-footed, five-toed form very similar to tracks from Utah, Colorado, and Arizona. In the Moab area *Brasilichnium* tracks range in width from less than 2 cm up to about 7 cm and resemble the footprints of small- or medium-sized dogs and the large ones resemble *Navahopus*. As discussed next, these tracks have been a source of confusion during the last generation because of the notion that they might be pterosaur tracks. To date, these "mammaloid tracks" (with a typical mammalian gait pattern) are known with certainty from Colorado, Utah, and Arizona, and possibly also from Nevada and California. They sometimes occur in conjunction with lizard-like tracks known as *Lacertipus* (meaning, lacertilian or lizard-shaped foot; figures 4.21 and 4.24) and small tracks of bipedal dinosaurs.

It is interesting to note that *Brasilichnium* tracks are very similar to *Laoporus* tracks from Permian sand dune deposits; in fact, many trackers have commented on the similarities. The two tracks are indeed hard to distinguish, despite being separated by five geological epochs—a gap of about eighty million years. [As just noted the last few million years of this gap may be filled by the recent discovery of mammal-like reptile tracks in the Moenave Formation.] The similarities between Permian and Jurassic occurrences of these tracks is further highlighted by the fact that the lizard-like trace *Dolichopodus* from the Permian Coconino Sandstone is almost identical to the Jurassic trackway *Lacertipus* (figure 4.24). Donald Baird, an ichnologist, has proposed that *Dolichopodus* represents the running trackway of *Laoporus*, which raises the interesting implication that *Lacertipus* is the trackway of a running *Brasilichnium* trackmaker.

For the purposes of this discussion we will presume that all four track types are different, and we will therefore preserve the different names. But we do stress that the issue needs further study in order to understand the range of variation seen in trackways and what this might imply about the locomotion of these animals. We also note that various arthropod traces attributed to spiders and scor-

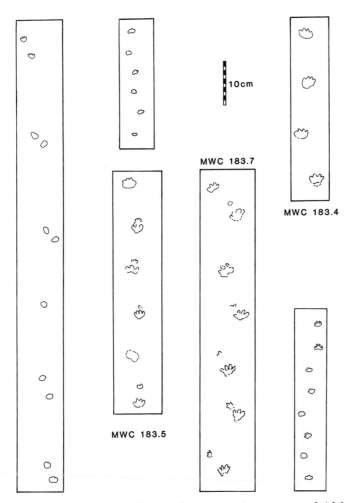

FIGURE 4.23 *Brasilichnium* trackways showing various mammaloid features—notably, wide feet and short toes. All examples are from the Navajo Formation, eastern Utah.

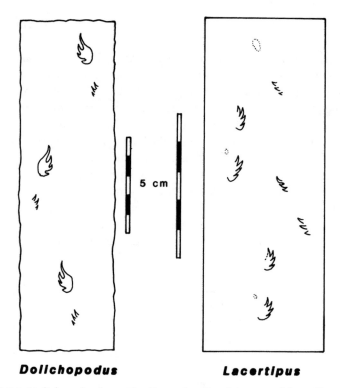

Dolichopodus   **Lacertipus**

FIGURE 4.24 *Dolichopodus* from the Coconino Sandstone and *Lacertipus* from the Navajo Sandstone are very similar trackways. Note that both have similar footprint orientations and step patterns. But the names are different because they were named by different trackers, who discovered them in formations of different ages.

pions (notably, *Octopodichnus*) are virtually identical in both the Permian and the Jurassic. This shows that animal communities inhabiting desert dune habitats during these widely separate time periods were very similar, or that they at least had similar foot structures.

The Pterosaur Tracks Dispute

Ever since pterosaurs, the famous flying reptiles of the Mesozoic, were discovered about two centuries ago, there has been debate about how they progressed on land. Most paleontologists have inferred that they moved on all fours, but finding trackways that could prove this inference has been difficult. The first purported pterosaur tracks, from Jurassic rocks in Europe, turned out to be a trackway of a horseshoe crab.

The dispute over pterosaur tracks was revived when William Stokes reported a set of tracks from the late Jurassic Morrison Formation (figure 4.25). The tracks in question were initially named *Pteraichnus saltwashensis*—meaning, pterosaur tracks from the Salt Wash Member of the Morrison Formation. Later, when the tracks were examined by paleontologists Kevin Padian and Paul Olsen, doubt was cast on this interpretation and it was suggested instead that the tracks were made by small crocodilians. It was further suggested that the *Pteraichnus* track-way was remarkably similar to the tracks of modern crocodiles and to the early Jurassic fossil trackway *Batrachopus* (figures 4.26 and 4.27). This seems to have been part of a trend in the 1980s to attribute many tracks to crocodilians. Thus, as we have seen, *Otozoum* (a prosauropod track), *Pteraichnus* (a pterosaur track) and *Batrachopus* (probably a true crocodilian track) have all been attributed to croco-diles at one time or another.

Before 1984, when Padian and Olsen disputed the pterosaur interpretation, Stokes had also attributed unusual tracks from the Navajo Formation to pterosaurs. After this date, however, it was assumed that these tracks were also crocodilian in origin, and at least one expert, Giuseppe Leonardi, concluded that

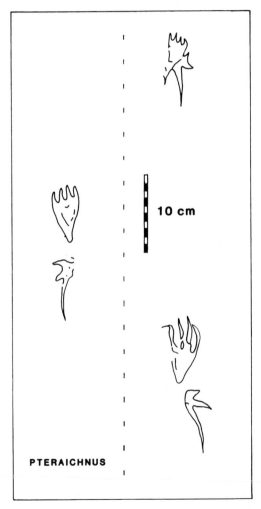

FIGURE 4.25 *Pteraichnus* tracks from the Morrison Formation of Utah represent the first true discovery of pterosaur tracks.   *After Stokes 1957.*

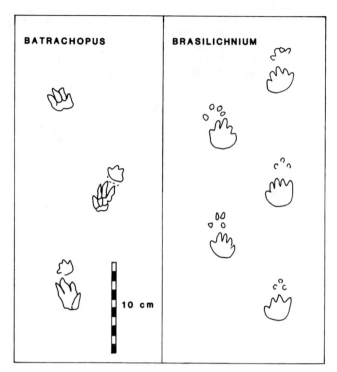

FIGURE 4.26 The crocodilian *Batrachopus* trackway can be distinguished from the mammaloid *Brasilichnium* trackway of the same time and region by the number and length of hind digit impressions. *Batrachopus after Padian and Olsen 1986; Brasilichnium after Chronic 1983.*

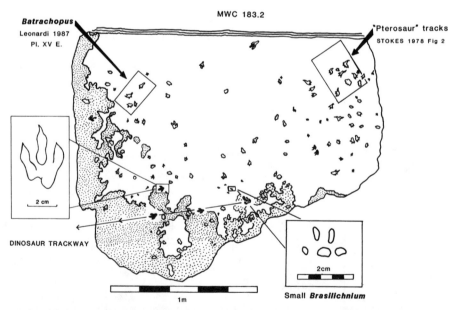

FIGURE 4.27 A slab of track-bearing sandstone from the Navajo Formation (companion to slab shown in Figure 4.22), showing tracks previously attributed to pterosaurs but now assigned to *Batrachopus*. Also shown are details of small mammaloid (*Brasilichnium*) and small dinosaur tracks.

they were best assigned to *Batrachopus* (figure 4.27). *Batrachopus* is reported in the western states only from the older Moenave Formation. Further study, however, revealed that the Navajo Formation tracks are not the same as *Batrachopus* and that they are neither crocodilian nor pterosaurian in origin. The majority are distinctly mammal-like in appearance and very similar to some tracks described from the Navajo Formation of Colorado and others described from Jurassic deposits in Brazil. In fact, several authors have commented on the distinctive symmetrical and broad shape of the foot (figures 4.21, 4.23, and 4.26), which is quite different from the elongate, asymmetrical *Batrachopus* footprint. Moreover, *Batrachopus* is four-toed on the hind foot, whereas well-preserved examples of *Brasilichnium* are five-toed.

Although the incorrect naming and interpretation of these Navajo tracks has resulted in much confusion, we are confident in drawing some fairly simple conclusions:

(1) There is no compelling evidence that any of the Navajo Formation tracks were made by pterosaurs.
(2) Mammal-like, probable tritylodont tracks are common in Navajo deposits (and in the Moenave) and also occur in similar age deposits in Brazil and probably in southern Africa. (Skeletal remains of tritylodonts have been found in early Jurassic deposits in the western United States and southern Africa.)
(3) The Navajo Formation footprint fauna is particularly diverse and useful for paleoecological reconstructions.

FIGURE 4.28 Typical slab with *Brasilichnium* trackways, Navajo Sandstone, Moab area.

What's in a Name?

One of the things that often frightens students away from paleontology is the plethora of long and apparently unpronounceable
names. Even though the jargon is actually no more difficult than
that of any other specialist science or foreign language, this does
not make it any more user-friendly for the student. As professional
paleontologists know only too well, sometimes the same fossil
(bone or track) has been given several different names by different
paleontologists working in different countries, often in different
generations or even in different centuries. This is the problem of
"provincial taxonomy," which in many cases is akin to the use of
different dialects, idioms, or languages in human society. The story
of the naming of the early Jurassic "mammaloid tracks" in North
and South America is a bizarre, but fascinating, example of the
complexities of "What's in a name?" Here we present a brief history of the discovery and naming of these mammaloid tracks.

1931 The famous German paleontologist Fredrick von Huene
reported probable mammaloid tracks from the early Jurassic
Botucatu Sandstone in Brazil. The specimens were not named at
that time or described in detail.

1951 Henry Faul and Wayne Roberts reported mammaloid tracks
from the early Jurassic (Navajo Sandstone) of Colorado. (No comparisons were made between these and the Brazilian specimens
and they were not named).

1957 Lee Stokes found probable crocodilian tracks in the Morrison
Formation of Utah; believing them to have been made by
pterosaurs, he named them *Pteraichnus*.

1970 Hartmut Haubold, another German paleontologist, named
the South American tracks *Tetrapodichnus* (meaning, tetrapod
trace), a name which he also applied to several other trackways
from various Permian and Triassic sites. (As pointed out later by
Giuseppe Leonardi, this name is not valid for the Brazilian specimen because it was not adequately described).

1971 Hartmut Haubold noticed that the mammaloid tracks from
Colorado described by Faul and Roberts had not been named, so he
named them *Bipedopus coloradensis* and *Semibipedopus meekerensis*.
He did not, however, study the original tracks and so he failed to
notice that they are not well preserved (there are no front footprints

(continued)

for *Bipedopus* and not all digit impressions are preserved). There is a double irony in the fact that Haubold applied the name *Bipedopus* to a track type that is evidently the same as those he had just assigned to *Tetrapodichnus* the previous year! In the same paper Haubold named lizard-like tracks, from the same Colorado beds, as *Lacertipus*.

1973–1974 Various authors first reported "pterosaur" tracks from the Navajo Formation. The implication was that these could also be called *Pteraichnus*.

1978 Stokes reported that mammaloid tracks are widespread in the Navajo Sandstone of Utah, Colorado, and Idaho.

1980–1981 After several years of study, Giuseppe Leonardi named mammaloid tracks from the early Jurassic Botucatu Formation of Brazil as *Brasilichnium elusivum* (meaning, elusive Brazilian tracks). His study was very thorough, and the trackways were described in detail, down to recognition of five-digit impressions in the hind footprints and a consistent pattern of front footprints in trackways. Even so, Leonardi did not mention similar tracks in North America, and he did not review the full history of the discovery and naming of mammaloid tracks in the early Jurassic.

1980 Don Baird named a trackway from the Navajo Formation as *Navahopus* and attributed it to a small prosauropod. Later it was suggested that this could simply be an example of a large mammaloid trackway.

1984 Kevin Padian and Paul Olsen reassessed *Pteraichnus* tracks from the Morrison Formation and concluded that they were made by crocodilians, not pterosaurs. This cast doubt on the purported pterosaur origin of tracks from the Navajo Formation.

1986 Paul Olsen and Kevin Padian reported the crocodilian track *Batrachopus* from the Moenave Formation of the early Jurassic, but they made no reference to the Navajo tracks. In the same year, however, they briefly examined some of the Navajo track slabs and identified *Batrachopus*, but did not publish their interpretation.

1987 After examining the same slabs (figure 4.27), Giuseppe Leonardi published a picture of certain tracks (possibly *Lacertipus* or *Brasilichnium*), assigning them to *Batrachopus*.

(continued)

1990–1993 Martin Lockley and his colleagues reexamined many of the mammaloid tracks and the purported pterosaur or *Batrachopus* tracksites in the Navajo sandstone. They published the first map of one of the controversial slabs (figure 4.21), indicating that it reveals the tracks of dinosaurs, mammal-like forms (similar to *Brasilichnium*) and a lacertilian form (similar to *Lacertipus*).

1993 Martin Lockley, Adrian Hunt, and Christian Meyer published a review of this history of confusion and suggested that *Navahopus* is remarkably similar to examples of certain large *Brasilichnium* tracks. They also reported that *Brasilichnium* is widespread in dune deposits in the Western Hemisphere (and possibly occurs also in southern Africa).

From the above it can be seen that the early Jurassic mammaloid tracks of the Western Hemisphere have been given a number of names, including *Tetrapodichnus, Bipedopus, Semibipedopus, Brasilichnium,* and possibly (depending on which tracks were looked at) *Pteraichnus, Batrachopus,* and *Navahopus*. We consider that all of the mammaloid tracks (first four names) are probably the same; only *Lacertipus* is possibly different but even this may be a function of gait. (*Pteraichnus* and *Batrachopus* are also different from the mammaloid tracks, but they evidently do not occur in the Navajo Formation). We conclude that the only adequate and valid description of any of these track types is that published by Giuseppe Leonardi for *Brasilichnium*. In our view, therefore, *Brasilichnium* should be the name used from here on out.

We can make one additional observation about tracks in the Navajo Formation. As we have discussed, most footprints other than *Brasilichnium* occur in flat-lying strata associated with playa lake deposits. Close examination of these layers reveals that the strata are contorted or heavily crinkled. Much of this irregularity or disturbance of the strata is due to trampling, or vertebrate "bioturbation" (disturbance of sediments by vertebrates). In recent years, as more examples of trampling have been recognized, the term "dinoturbation" has been coined, and it is now common parlance among many geologists working on Mesozoic terrestrial strata in the West.

In many instances, the proof that crinkled strata owes to dinoturbation comes in finding surfaces covered with tracks. Such surfaces may be very irregular or bioturbated with tracks showing varying degrees of good and bad preservation. Some tracks will overlap and distort others, whereas some layers will be so churned or ploughed up by trampling as to render any individual tracks unrecognizable. In such cases, even though the beds may "look" trampled it is often hard to prove trampling or demonstrate which animals were responsible. Several examples of trampled beds are known from Navajo playa deposits (for example, the moderately trampled bed in figure 4.13). The "dinoturbation" phenomenon is widespread in later Mesozoic deposits, as well.

Early Jurassic Climates and Biogeography

Through much of the late Paleozoic era and into the early Mesozoic the world's continents were joined into a large supercontinent known as Pangaea II (meaning, "all earth" or "wide earth"; figure 4.29). The northern part of this continent, comprising what is now Europe, Asia, and North America, is referred to as Laurasia; the southern part—South America, Africa, Australia, and the Antarctic—is Gondwanaland. Because all the continents were joined, animals could move freely across most of the earth's land surface. This created a situation where few if any groups were isolated and able to evolve completely indepen-

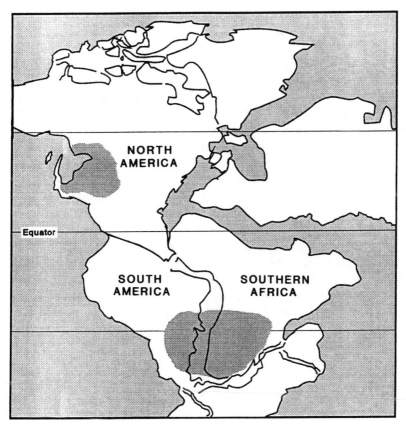

FIGURE 4.29 Pangaea II during the early Jurassic. Desert regions are marked by stipple.

dently. Animals and animal communities were pandemic (widespread) rather than endemic (localized) in their distribution.

Not surprisingly, track types before the breakup of Pangaea II are similar on a global scale, thus allowing trackers to correlate footprints now found on continents that have long since separated. This, in turn, has fostered the science of palichnostratigraphy, in which rock formations that are geographically distant from one another can be correlated in time through the use of ancient tracks. Palichnostratigraphy is particularly useful in studies of Permian through early Jurassic rocks, and in trying to assign ages to rocks that have few other fossils.

Animal communities, however, were also influenced by environmental conditions, or what ecologists call "limiting factors," such as climate and food resources. For example, some creatures preferred arid environments over humid ones. Thus, it is not just the physical separation or joining of continents that leads to isolation or mixing of animals. The constraints imposed by climate, food availability, and other limiting factors within a particular continental area are important, too. As a result, tracks from desert dune deposits tend to be different from those found in deposits representing wetter climates, even though the tracks may be exactly the same age. It is thus essential to compare tracks from similar environments when undertaking age comparisons or palichnostratigraphic studies.

The *Brasilichnium* dunefield ichnofacies of the early Jurassic provides an excellent example of this problem. This ichnofacies is very distinctive; the ichnofacies with the closest resemblance is that of the Permian *Laoporus*, which is about 70–80 million years older. The *Brasilichnium* ichnofacies should probably not, therefore, be compared with other non-dunefield track assemblages of the early Jurassic, such as those that contain *Eubrontes*, *Anomoepus*, and *Otozoum*.

The supercontinent configuration of the late Paleozoic and early Mesozoic played an important role in global climate patterns. Generally, large continents create so-called continental climates, especially in their interiors. Dry conditions tend to prevail with harsh extremes of heat and cold between summer and winter and between day and night. The widespread distribution of ancient dune sediments in Carboniferous, Permian, Triassic, and Jurassic times—especially in North America—supports this generalization. This long sequence of time was not, of course, uniformly dry and arid; there were both humid times and humid regions, and the animal communities evolved and adapted accordingly. For example, late Triassic deposits of North America mainly indicate humid (monsoonal) conditions, with only occasional, short-lived dry spells. During the early Jurassic and into the middle Jurassic, desert conditions tended to prevail in this same geographic region, with periodic wetter phases. It is against this background that we begin our discussion of the animal communities adapted to desert conditions.

Although deserts are usually regarded as inhospitable to diverse life, the fossil track record indicates a surprisingly varied fauna. It is crucial to remember, however, that most tracks—even in desert environments—are made and preserved on moist substrates. This means near water, and water attracts life. Many of the tracks discovered in dunefield or other arid deposits were made in and around oases or watering holes. Even though Jurassic desert deposits are replete with tracks, tens of millions of years are represented by the strata. Thus the hundreds of known trackways represent, on average, the proven activity of only a few animals every million years.

These geological considerations notwithstanding, we can infer that the tracks represent a part of the indigenous animal communities that inhabited desert or arid landscapes. The overall picture appears to be one of vertebrate faunas dominated by small- to medium-sized bipedal dinosaurs.

Were Dinosaurs Mostly Meat-Eaters?

It is known that the tracks of bipedal, theropod dinosaurs often predominate in footprint assemblages. Although this might be taken as evidence of large populations of carnivores and insectivores, there is another possible explanation. Meat-eating dinosaurs may have been more active, rather like many bird species. They may have patrolled oases and water courses searching for potential prey species, including aquatic crustaceans and other "small fry." Given that many of the trackmakers were no larger than an emu and many more were chicken-sized, the bird analogy is worth considering. Birds today are often very active, leaving many tracks in shoreline habitats.

Some paleontologists have pointed out that not all ancient animal communities were organized like modern mammal communities, where a small number of mammalian carnivores rely on a larger number of herbivorous mammalian prey species. This classic idea of a food chain (more properly, an "ecological pyramid") derives from the assumption that vegetation (primary producers) sustains a large number of herbivores (primary consumers), which in turn support a few top predators or carnivores (secondary consumers). Figure 4.30 is a sketch of a modern ecological pyramid. But what happens in desert and arid habitats where vegetation is sparse or absent? In such environments the local character of the food chain may be quite different. Small vertebrates, like rodents, birds, and lizards, may rely on insects and other arthropods for sustenance, which may be sustained by microscopic detritus and other arthropods. The insectivorous small vertebrates may in turn be preyed upon by larger animals in a carnivore-eat-carnivore food chain.

In Wingate, Moenave, and Kayenta deposits, footprint assemblages dominated by bipedal theropods do indeed suggest a carnivore-dominated vertebrate community. In the Navajo Formation, however, the picture is rather different. A diverse animal community is represented (figure 4.31), showing differences between the dune track assemblages (mainly *Brasilichnium* and small theropods) and interdune track assemblages (*Otozoum* and both large and small theropods). Although carnivores were present in the form of theropods, prosauropod tracks indicate the presence of herbivores. Skeletal fossils of the mammal-like reptiles, such as the tritylodonts, indicate that they were probably herbivorous also, or perhaps omnivorous. As many playa deposits contain traces of arthropods—perhaps made by crayfish-like crustaceans—it seems probable that dinosaurs and other vertebrates would have availed themselves of such food resources.

The presence of herbivorous prosauropods indicates that vegetation existed (at least in patches) in some early Jurassic desert environments. This conclusion

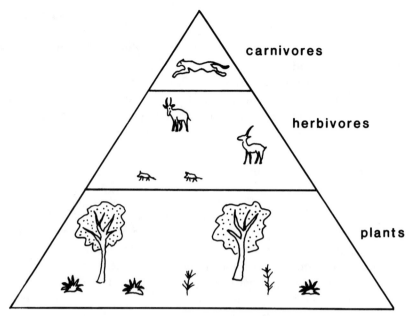

FIGURE 4.30 An ecological pyramid showing the relative proportions of primary producers (plants), primary consumers (herbivores), and secondary consumers (carnivores).

is supported by the discovery of fossil wood in association with oases or playa lake deposits. Even though prosauropods may have favored dry or semi-arid habitats, they could not have survived indefinitely in hyper-arid settings.

Overall, the early Jurassic track record in the western United States is clearly useful in reconstructing the animal communities of that time. But it is not yet clear why the older Wingate, Moenave, and Kayenta footprint faunas differ from the younger and more diverse Navajo Formation track assemblages, so more research is necessary. The diversity of the Navajo may owe to several million years of additional evolution and dinosaur radiation, to changing environmental conditions that allowed more diverse animal communities to become established in the region, or to artifacts in the preservation and collection of track data. In any event, the footprint record shows that early Jurassic desert environments were by no means devoid of life.

The Mid-Jurassic and the Moab Megatracksite

The middle Jurassic record of *skeletal* remains is pitifully sparse in the western United States and in North America in general. Only one vertebrate fossil has thus far been discovered—a small crocodilian from the Entrada Sandstone in eastern Utah. However, small dinosaur tracks are known from the Carmel Formation and larger dinosaur tracks appear in the Entrada Sandstone. The Sundance and Summerville formations (which represent the coastal plain of an encroaching sea) contain pterosaur tracks and may in part be late Jurassic. In each of these three middle Jurassic formations, therefore, the track assemblages are distinct. The top layer of the Entrada, moreover, contains so many tracks at its upper boundary with the Summerville that it qualifies as one of the grand "megatracksites" of the West. [In places the upper part of the Summerville is referred to as the Tidwell; a stratigraphic classification that is currently controversial: see figure 4.1 for illustration.]

The middle Jurassic track record appears to be restricted to mainly footprints of three-toed dinosaurs of various sizes, in most cases probably theropods. Some enigmatic, possible crocodilian tracks are known from the Carmel Formation, and pterosaur tracks are now known from the Summerville and Sundance formations, but dinosaurs are clearly the stars of middle Jurassic strata. Overall, the vast majority of middle Jurassic tracks have only just been discovered and adequately interpreted, so the study of middle Jurassic dinosaur communities and paleoecology is still at a very early stage. Despite this limitation, knowledge of middle Jurassic tracks has improved substantially in recent years and provides valuable insight into the vertebrate communities that existed in North America.

DINOSAUR I

DINOSAUR II

DINOSAUR III

MAMMAL–LIKE REPTILE

LIZARD

SCORPION

FIGURE 4.31 The diverse animal community of the Navajo Formation, reconstructed from trackways found in association with playa or oasis deposits.

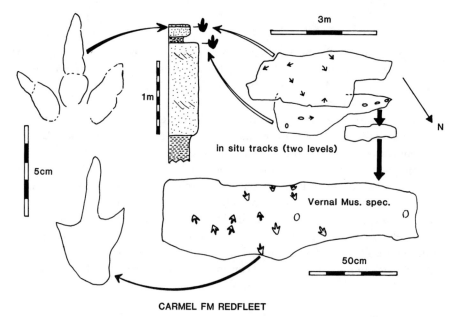

CARMEL FM REDFLEET

FIGURE 4.32 Tracks from the Jurassic Carmel Formation, Redfleet Reservoir area, northeastern Utah. The lower slab is on display at the Utah Field House of Natural History in Vernal, Utah.

The Carmel Formation represents a break or hiatus in the accumulation of desert dune deposits, as Carmel rocks are primarily marine and coastal plain deposits. During this time seas periodically encroached on the continent from the west and northwest, turning sand dunes into low-lying sandflats, mudflats, and poor soils. This encroachment did not occur or get recorded in all areas. This is either because the sea never reached particular areas or because the marine deposits were subsequently removed by erosion, leaving a gap or unconformity in the rock record. For example, in many regions, especially eastern Utah, the early Jurassic Navajo Formation is directly overlain by the Entrada Sandstone, whereas elsewhere, as in northeastern and central Utah, the Carmel Formation occurs between the Navajo and the Entrada.

At least two dinosaur tracksite areas are known in the Carmel, both in Utah. The first is at Redfleet Reservoir north of Vernal, and the second is just south of Dinosaur National Monument, near Jensen. The Redfleet site has yielded several dozen small tracks indicative of chicken-sized dinosaurs (figure 4.32). Some of these are on display at the Utah Field House Museum in Vernal. The Jensen site has yielded a greater range of track sizes, including emu-sized tracks, some elongate scrape marks that are reminiscent of traces sometimes attributed to swimming animals, and at least one slender-toed, bird-like track (figure 4.33). It is probable that all the three-toed tracks (figure 4.34) were made by theropod dinosaurs. The majority of footprints have been assigned the name *Carmelopodus* (meaning track from the Carmel formation). The elongate scrape marks resemble certain Cretaceous traces that have been attributed to crocodilians or pterosaurs.

The Entrada is another eolian, sand sea deposit that has traditionally been regarded as devoid of paleontological evidence. Although bones are restricted to one occurrence, recent discoveries reveal the presence of tracks at about thirty sites in eastern Utah. All the tracks known to date are those of medium to large carnivorous dinosaurs.

Almost all reported Entrada Sandstone tracksites are associated with the very top of the formation in the vicinity of Arches National Park, near Moab, Utah. These sites, however, are not isolated concentrations or patches of tracks in different layers or strata. They are all part of a single vast expanse of tracks cov-

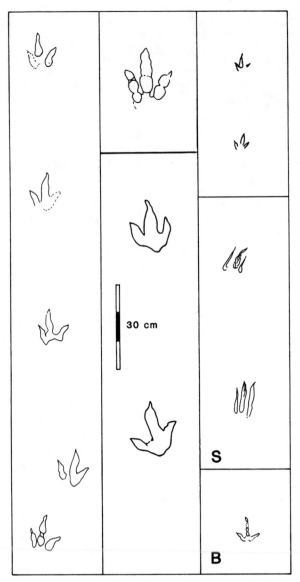

FIGURE 4.33 Tridactyl tracks from the Carmel Formation include *Carmelopodus* (theropodan), probable swim traces (S) and bird-like track (B). After Lockley et al. (1998).

FIGURE 4.34 Tridactyl (theropod) track from the Carmel Formation, near Dinosaur National Monument, has been named *Carmelopodus*.

ering an area of over 300 square miles (about 1000 square kilometers) and occurring only in the uppermost (youngest) layer of the formation, where it passes up into the overlying strata of the Summerville Formation. Figure 4.35 maps these Entrada sites, which may be late Jurassic in age.

Such extensive track-bearing layers have been referred to as "megatracksites." Megatracksites in general, by definition, are regionally extensive. Significantly, they tend to be associated with the tops of formations, that is, at the boundaries or contacts between formations. In the case of the Moab megatracksite this can lead to some confusion. Were the tracks made during the final stages of Entrada time or at the beginning of Summerville deposition in this area? There is even a third possibility—that the tracks were made after the accumulation of Entrada Sandstone but before the beginning of Summerville deposition. In other words, the tracks represent dinosaur activity during a hiatus or break in sediment accumulation.

Such breaks in sediment deposition are well known to geologists and are termed *unconformities* (where a deposit rests without conformity or continuity on the deposit below it). Tracks are often associated with such unconformities. This is a paradox, in a way, almost a contradiction: a record of dinosaurs where the rock is missing.

Regardless of the complexities involved in the geological sequence of formations in this region, we can state that the Moab megatracksite is associated with some type of unconformity or hiatus late in the middle to late Jurassic. We also know that this is the oldest megatracksite currently known and that, like the Cretaceous examples to be discussed in chapter 5, the track-bearing layers represent some type of low-lying coastal plain environment, probably close to sea level. In such coastal settings, broad low-lying regions are subject to inundation by the sea or to wetting by proximity to coastal lagoons and waterways. Overall, a megatracksite is a sign that extensive areas were good for trackmaking.

Estimates suggest that there are literally billions of tracks in the Moab megatracksite. One track per square meter equals one million per square kilometer. In more heavily trampled or dinoturbated areas, ten tracks per square meter equals ten million per square kilometer, and so on. So, in this thousand-square-kilometer area there are between one billion and ten billion tracks, based on an average density estimate of between one and ten per square meter. Bearing in mind the unconformity phenomenon, such astronomical footprints numbers and trampling are not necessarily attributable to a huge rise in the population of dinosaurs. One must be careful not to jump to a biological conclusion and postulate a population explosion. Instead, the track zone results largely from the long span of time during which the substrate was available for trackmaking, and the simple fact that conditions were suitable for the preservation of this zone.

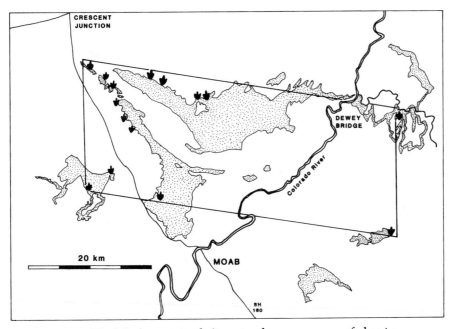

FIGURE 4.35 The Moab megatracksite extends over an area of about a thousand square kilometers. At least thirty individual tracksites that have been mapped are known to occur on the same surface.

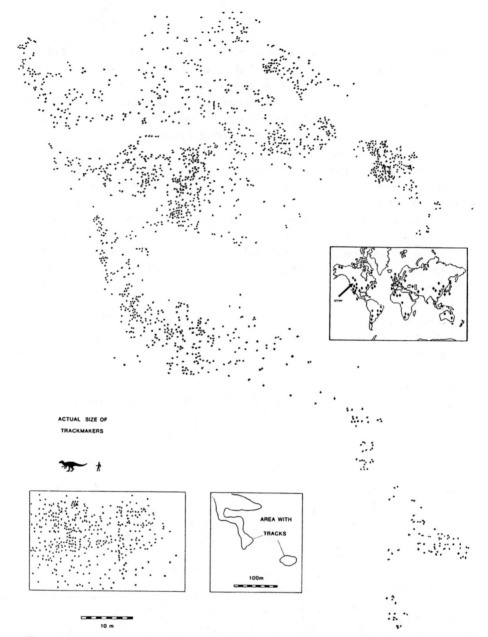

ACTUAL SIZE OF
TRACKMAKERS

AREA WITH
TRACKS

100m

10 m

FIGURE 4.36 A large tracksite, near Moab, is known as the "Stomping Ground." It forms part of the megatracksite complex at the Entrada-Summerville boundary. Each dot represents one track. Only theropod tracks are represented at this site. *After Lockley et al. 1991.*

Whatever the cause, the large number of tracksites and tracks available for study represents an explosion in available data for paleontologists. Figure 4.36 gives a visual indication of the wealth of tracks available from just one part of the mega-tracksite, known as the "Stomping Ground." Here we see about 2,300 tracks in an area of about two acres.

Like the majority of Carmel Formation footprints, tracks from the Entrada-Summerville boundary sequence are all attributable to three-toed dinosaurs, probably theropods (figure 4.37). However, unlike the Carmel track assemblage, the Entrada-Summerville tracks are all relatively large, and have been named *Megalosauripus* (figure 4.38). The vast majority fall in the range of 30–45 cm (foot length), and in this respect contrast quite markedly with the assemblage of diminutive tracks found in the Carmel. This raises the intriguing question—why are all the tracks in one formation small, and all those in another formation large? This question may never be answered in full, but it does suggest that we can at least infer different dinosaur communities inhabiting different environments.

One track-bearing bed about five meters below the main Entrada-Summerville boundary level contains an assemblage of smaller theropod tracks, with foot lengths of about 15 cm. This noticeable size difference again seems to suggest that a different animal community was represented during Entrada times, and within the dunefield environment, compared to the fauna that was

FIGURE 4.37 A typical theropod trackway from the Moab megatracksite. *After Lockley 1991.*

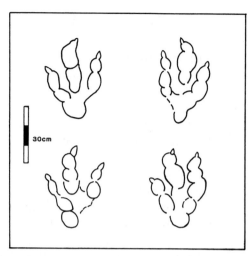

FIGURE 4.38 Large theropod tracks representative of the Moab megatracksite have been assigned to the ichnogenus *Megalosauripus* (Lockley et al., 1996).

represented at the end of Entrada time as the dunefield environments were replaced by Summerville coastal plain environments. Remember that there are no dinosaur skeletal remains known from the Entrada, or the Carmel, so the tracks are all we have to go on in reconstructing the paleoecology.

The distribution of tracks at the inter-formation boundary cannot be attributed solely to the presence of many theropod dinosaurs. Instead it is probably a preservational phenomenon associated with the unconformity. We infer that tracks were made and preserved as the sea encroached from the west, moving further east, wetting strips of coastal plain shoreline and leading to the lateral accumulation of a thin, but widespread, track-bearing zone.

Our former colleague, John Foster, has recently discovered *Megalosauripus*-like tracks at the top of Entrada Formation near Escalante in south central Utah, 135 miles southwest of Moab. Could it be that he has discovered an extension of the Moab megatracksite? Such important finds suggest the potential to predict the location of other sites at these levels elsewhere in the region, and reveal the huge extent of such megatracksites.

The Real Pterosaur Tracks Story

The Summerville Formation represents a widespread low-lying coastal plain of a sea that encroached into eastern Utah and western Colorado toward the end of the middle Jurassic and the beginning of the late Jurassic. In places it contains gypsum, indicating a salty lagoonal setting. It has produced no fossils of terrestrial vertebrates, but it has produced some tracks. To the west, marine fossils are known, and these have been used to date the Summerville Formation as the late part of the middle Jurassic. The discovery of tracks within the Summerville itself (not part of the megatracksite at its contract with the Entrada) is a significant addition to paleontological knowledge because it provides the only evidence of vertebrates from this formation. The tracks (figure

4.39) also appear to be quite unusual for western North America as they are best attributed to pterosaurs (*Pteraichnus*).

We came to this conclusion in a roundabout way—something of a wild pterosaur chase, one might say. Remember that Padian and Olsen disputed the interpretation of *Pteraichnus* tracks as pterosaurian, effectively concluding that no proven pterosaur tracks exist anywhere in the world. We had gone along with this conventional wisdom until very recently. When we first examined the Summerville tracks, discovered by our colleague John Robinson, we knew they were *Pteraichnus*-like, but why did the trackway show only the front footprints? Could it have been made by a swimming crocodile? This pattern of trackways with only front footprints is known for swimming turtles from late Jurassic deposits in Europe. So we initially opted for a turtle interpretation.

Perhaps the key to the puzzle is in trackways of *Pteraichnus* from the Sundance Formation of Wyoming—a deposit equivalent to the Summerville of

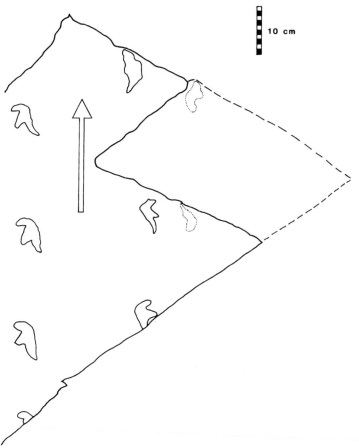

FIGURE 4.39 A trackway from the Summerville Formation of eastern Utah has been attributed to a pterosaur. The three-toed manus track with its sideways orientation is typical of *Pteraichnus*.

Utah. In 1977 Terry Logue reported pterosaur tracks from the Sundance; but after 1984 most paleontologists assumed they had been reinterpreted as crocodilian. Logue, however, still believed in the pterosaur tracks interpretation and said so in print as recently as 1994. These Wyoming trackways are abundant and reveal many similarities to Stokes's original *Pteraichnus* discovery (figure 4.40). They also exhibit very good details of the hind foot shape that match the pterosaur foot skeletons very closely (figure 4.41). (Differences in the length of the longest front foot digit and the outward rotation of the hind foot suggest a different species of *Pteraichnus*). They also show front footprints that are deeper than the hind foot-

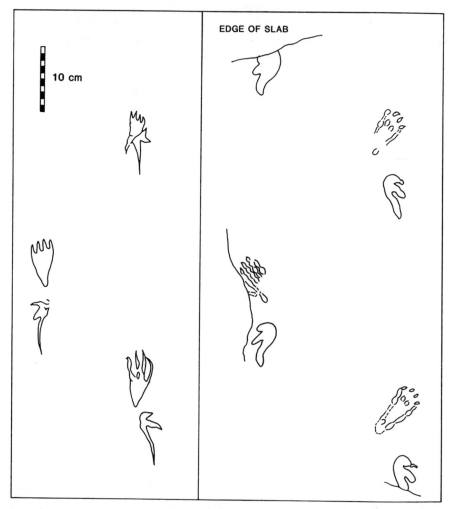

FIGURE **4.40** Comparison between the type specimen of *Pteraichnus* from the Morrison Formation of Arizona on the left (after Stokes, 1957) and the new trackway from the Sundance Formation of Wyoming on the right. Note the better-preserved pad impressions and the wider trackway (= lower pace angulation) in the Sundance specimen.

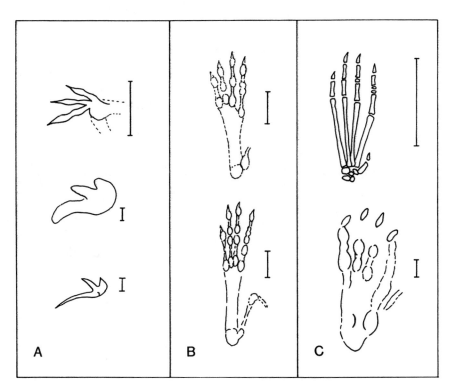

FIGURE 4.41 Details of *Pteraichnus* manus and pes tracks, compared with hypothetical reconstructions. A: hypothetical manus track, after Unwin 1989 (top); manus track from Sundance Formation (middle); *Pteraichnus* type trackway (bottom). B: hypothetical pterosaur pes tracks based on *Rhamorphorhynchus* (top) and *Pterodactylus* (bottom), after Unwin 1989. C: *Pterodactylus* pes, after Wellnhofer 1991 (top); composite pterosaur pes track, from Sundance Formation, Wyoming (bottom).

prints. This is an important clue to why the Summerville trackway only shows front footprints. As the sand that makes the layer above the footprints was washed in, it scoured away the shallow hind tracks but filled in and preserved the front footprints.

We now know of four pterosaur tracksites, two in the Summerville-Sundance marine coastal deposits and two, including Stokes's original report, said to be immediately above this level in the lower part of the Morrison Formation (our own investigations at Stokes' original *Pteraichnus* locality in Arizona have revealed tracks in the top of the Summerville, not in the Morrison). This distribution suggests a compelling story because almost everyone agrees that most pterosaurs were fisher folk like many large seabirds. Moreover, based on skeletal evidence, they began to become very abundant and diverse in the late Jurassic. Their tracks in the western United States, not surprisingly, coincide with the encroachment of the sea and the end of widespread desert conditions in the region (figure 4.42).

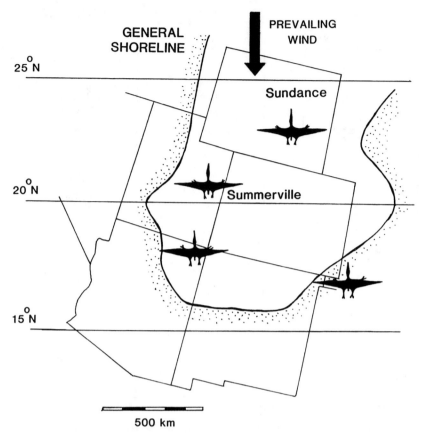

FIGURE 4.42 Paleogeography of the marine embayment of the middle to late Jurassic, showing the occurrence of four different *Pteraichnus* tracksites in the succession of rocks that run from the Summerville Formation into the Sundance and finally into the basal sections of the Morrison Formation. *After Kocurek and Dott 1983.*

The identification of *Pteraichnus* as pterosaurian, not crocodilian, also goes a long way toward solving a century-old debate about how pterosaurs walked on land. The questions that have been asked are: Were they quadrupeds or bipeds? Were they semi-erect or erect walkers? Were they plantigrade (flat-footed) or digitigrade (toe-walkers)?

Although most have opted for a quadrupedal, semi-erect, flat-footed form (figure 4.43), others—including Kevin Padian—favor an erect, bipedal, digitigrade animal. The trackway evidence clearly shows that the quadrupedal interpretation is correct, at least for these trackmakers. The tracks also show that, when in contact with the ground, pterosaur wing claws rotated sideways rather than pointing forward, as in previous reconstructions. In summary then, we conclude that pterosaur trackways do exist and that they indicate quadrupedal progression on land, at least for some species. The tracks, like the skeletal remains, also indicate a coastal habitat for some pterosaurs.

One of the most recent and interesting Summerville finds was the discovery, by geologist John Robinson, of several sauropod footprint casts showing skin impressions (figure 4.44). (A cast is a preserved infilling of an impression.) The sauropod casts were found near the contact between the Tidwell Member of the Summerville and the overlying Saltwash Member of the Morrison Formation, near Bullfrog, Utah, on the north side of Lake Powell. This find represents the first report of skin impressions from a sauropod footprint anywhere in the world. Although the skin impressions imply excellent preservation, the overall shape of the foot casts is a little puzzling. As shown in figure 4.44, the specimens contain what appear to be small toe claws. These are preserved at the end of two long prominent grooves that run down the sides of the casts. The claw marks are probably best interpreted as the anterior claws on the hind foot (pes), in which case the pes would be by far the smallest sauropod hind foot track recorded in the Morrison, and in all of North America (only about 35 cm long by 20 cm wide).

Part of the reason that these particular tracks appear so unusual is that the tracks are not imprints but natural casts. Almost all known sauropod tracks in North America, and in most of the rest of the world, are relatively shallow impressions in limestone or sandstone that were later filled in by shale or other softer material. But in the unusual case of the casts discovered near Bullfrog,

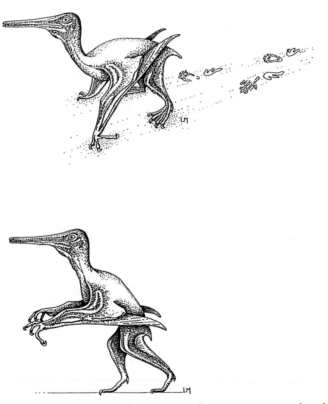

FIGURE 4.43 Reconstruction of quadrupedal pterosaurian trackmaker, showing manus digits rotated to the position recorded in the rocks, with a bipedal pterosaur for comparison. *After Wellnhofer 1991.*

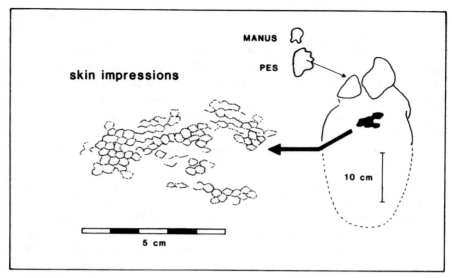

FIGURE 4.44 Sauropod track with skin impressions, Tidwell Member of the Summerville Formation, Bullfrog area, Utah. *After Lockley et al. 1992.*

Utah, a deep track was made in mud and then filled in by sand that became a more resistant sandstone. While ichnologists are familiar with the appearance of shallow brontosaur tracks made on relatively firm surfaces, none of us has much experience of deep tracks made in soft, very malleable substrates. Deeper tracks in soft substrates have the potential to preserve considerable detail of the foot shape and claw and skin impressions, as is evidently the case here.

Morrison Formation Tracks of the Late Jurassic: The Golden Age of Brontosaurs

To date, almost all known late Jurassic tracksites (including a possible pterosaur tracksite) in the western states are found in the Morrison Formation, which is famous for the skeletal remains of dinosaurs like *Apatosaurus* (also known as *Brontosaurus*), *Allosaurus*, and *Stegosaurus*. These and other dinosaur bones can be admired in situ at Dinosaur National Monument. For this reason skeletal remains have overshadowed the track record, and few people think of footprints when they think of the Morrison Formation.

Nevertheless, at least thirty fossil footprint sites are known from the Morrison Formation. Most reveal evidence of the activity of dinosaurs, mainly theropods and sauropods, but a few reveal evidence of the activity of pterosaurs and other reptiles. Some dinosaur groups that are relatively common as body fossils, notably the stegosaurs, have, as yet, little or no known track record.

Unlike the Moab megatracksite of the middle Jurassic, most of the thirty known tracksites of the late Jurassic reveal only a few footprints at each location in strata at different levels in the formation. One such example is the site in the Valley City area between Moab Airport and Crescent Junction in eastern Utah. Another example of a small site was discovered near Fruita, Colorado. Both are in the Salt Wash Member, near the base of the Morrison.

At the Valley City site are five trackways in a small area—one of a large sauropod and four of theropods (figure 4.45). This site is interesting for two reasons. First, the sauropod was changing direction, turning to the right, a phenomenon rarely recorded in trackways. Second, one of the theropods was taking alternate long and short steps, perhaps limping. We do not know the reasons for these unusual behaviors, just that they happened. This site is deemed so interesting that it has been furnished with signs and interpretative brochures and is accessible to the public.

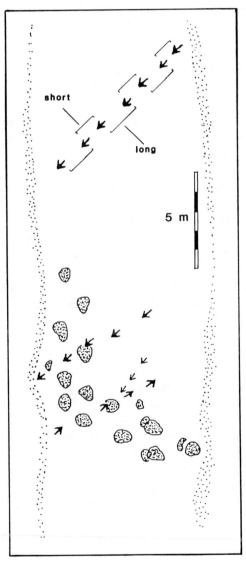

FIGURE 4.45 Map of the Valley City tracksite, Morrison Formation, eastern Utah. Note the turning brontosaur trackway (bottom) and the alternating long and short steps of the isolated theropod trackway (top).

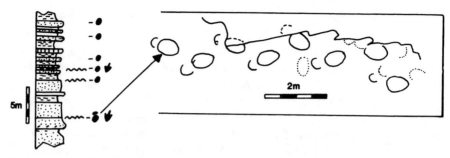

FIGURE 4.46 The Rancho Del Rio dinosaur tracksite in Colorado reveals at least six bedding surfaces in the Morrison Formation that contain tracks, some with extensive dinoturbation. The tracks are of theropods and sauropods. A surface with sauropod tracks is depicted here. Note the typical oval hind footprints and smaller semicircular front footprints. *After Lockley 1986.*

The Fruita site is only one of several small tracksites in this general region. Most reveal only a few three-toed tracks, though one site preserves the tracks of a four-toed, lizard-like creature.

Sometimes small sites such as these are not documented (that is, described and published in a scientific journal) because the isolated track occurrences are not considered of any great significance, or because other larger and more important tracksites are a greater priority for study. Ultimately, such neglect is detrimental to the science of ichnology because the data are then only available to those who happen to know about and can visit the site. The data are not, therefore, integrated into the overall data base for a given area or time interval; worse, the site itself may be forgotten and eroded as the years pass. As we discussed earlier, a census of the relative numbers of trackway types from all localities in a given formation may provide a useful perspective on the composition of faunas. Such track information has merit in its own right, but it can also prove useful for comparison with the record of skeletal remains. (We will undertake such a comparison at the end of this chapter.)

Only two Morrison tracksites qualify as large sites. One is at Rancho Del Rio, along the Colorado River near the town of State Bridge in central Colorado. The other is along the Purgatoire River in southeastern Colorado. The Rancho Del Rio site reveals theropod and sauropod tracks at several different levels near the base of the formation. It contains evidence of trampling of the substrate by these dinosaurs. Figure 4.46, for example, shows at least six track-bearing levels.

Very recently we ourselves discovered another site near State Bridge. This site is interesting because in addition to the tracks of small- to medium-sized theropod dinosaurs, it includes one trackway that resembles that of a large heron, with widely splayed toe impressions (figure 4.47).

These small sites, though important, are dwarfed by the late Jurassic tracksite in the Purgatoire Valley of southeastern Colorado. The Purgatoire site is now famous as one of the world's largest and most spectacular dinosaur footprint assemblages (figure 4.48). It is accessible for viewing by tourists and amateurs, as well as by professionals. Over 1,300 dinosaur tracks have been mapped in a single layer (level 2 in figure 4.49), and tracks occur at three additional levels.

The footprints indicate an animal community dominated by sauropods and theropods of different sizes (figures 4.50, 4.51, 4.52, and 7.3).

The huge extent of the Purgatoire site, which is about 400 meters in its longest dimension, has made it particularly useful for reconstructing the ancient environment in which dinosaurs of the Morrison thrived. Along with the tracks can be found fossils of plants, algae, snails, clams, crustaceans, and fish—all indicative of a freshwater or slightly alkaline lake. This ancient habitat has been dubbed "Dinosaur Lake" and the tracks show the dinosaur community that frequented its shoreline.

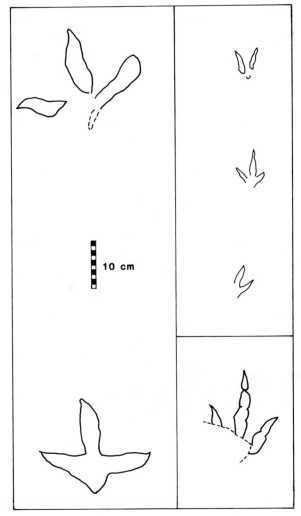

FIGURE 4.47 The State Bridge tracksite of Colorado reveals an assortment of small- to medium-sized theropod tracks, including one type in which the toes are so splayed that it resembles footprints of a large heron-like bird.

FIGURE 4.48 Helicopter photograph of Purgatoire Valley dinosaur tracksite, southeastern Colorado. Note prominent brontosaur trackways.

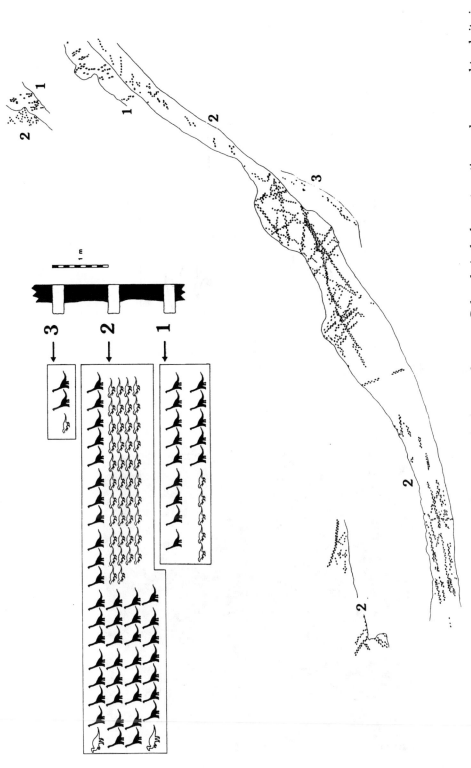

FIGURE 4.49 The Purgatoire dinosaur tracksite ("Dinosaur Lake") in southeastern Colorado is the largest continuously mapped tracksite in North America in terms of area—extending for more than 350 meters. Mapped sauropod and theropod tracks occur at three different levels. The relative abundances of the two track types in each level are shown using a silhouette for each individual trackmaker. *After Lockley et al. 1986.*

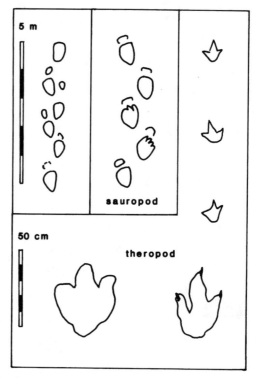

FIGURE 4.50 Tracks representative of the Purgatoire tracksite.

FIGURE 4.51 Photograph of a small theropod track from the Purgatoire tracksite.

The site has revealed fascinating evidence of gregarious behavior. One group of five brontosaur trackways (figure 4.53) shows that small animals walked parallel to a lake shoreline. These five trackways, moreover, were presumably made by subadults, as there are no species of small sauropods known from body fossils. The presence of subadult sauropod individuals is noteworthy because most

FIGURE 4.52 Sauropod tracks at the Purgatoire tracksite, as photographed in 1938.

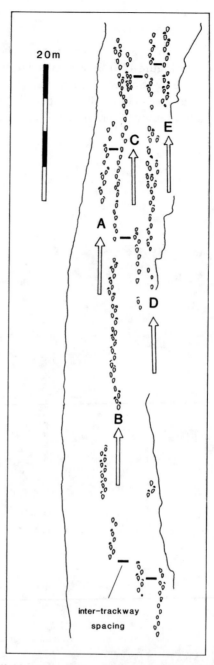

FIGURE 4.53 Parallel brontosaur trackways at the Purgatoire site strongly suggest gregarious behavior. *After Lockley and Prince 1998; Lockley 1991.*

sauropod body fossils represent large individuals. Thus it appears that the sauropod body fossil record in the Morrison Formation may be biased toward the preservation of large individuals, whereas the track record shows a wider, more representative range of sizes.

In some layers at Dinosaur Lake, track density is so high that the layers have been pulverized by trampling. This phenomenon of dinoturbation is quite widespread in late Mesozoic deposits where large gregarious dinosaurs had been abundant or inclined to repeatedly frequent wet substrates. Trampling or trampled layers first became common in what is now the western United States in the early Jurassic, particularly in association with localized playa lake deposits. Heavily trampled layers are also common in parts of the Entrada megatracksite complex near Moab, and they remain common in many Cretaceous deposits. Thus the Morrison examples are fairly typical of the disturbance produced by large dinosaurs in Jurassic and Cretaceous times. It was not just the sediment that was disturbed; at one level clams have been crushed and destroyed by the impact of brontosaur feet (figure 4.54) and plant stems have been flattened into the limey lakeshore sediments.

Adding to his good fortune, John Robinson—discoverer of pterosaur tracks and sauropod skin impressions—proceeded to make a second find of sauropod casts in the Saltwash Member of the Morrison Formation in the Bullfrog area of Utah. Robinson discovered two large, well-preserved sauropod track casts (figure 4.55) that protrude from beneath a sandstone ledge in the Saltwash Member of the Morrison Formation. These tracks, originally made in mud (the mudstone has eroded away, leaving the sandstone hanging as a ledge), show excellent details of the toes of the sauropod hind feet, including their claws. It is surprising that although late Jurassic sauropod trackways are known from several locations

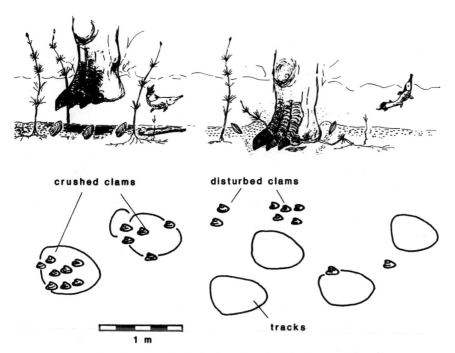

FIGURE 4.54 A trackway that includes clam fossils. Some of the clams appear to have been crushed by brontosaur trampling.
After Lockley et al. 1986; Lockley 1991.

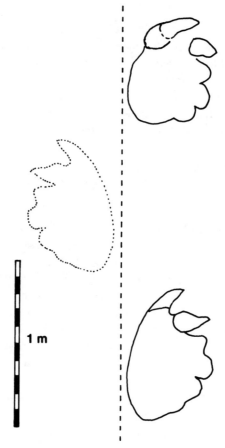

FIGURE 4.55 Two consecutive right footprint casts of a large sauropod from the Salt Wash Member of the Morrison Formation, near Bullfrog, Utah. The left track is inferred.

around the world, few have well-developed claw impressions. Thus the Bullfrog sites reveal the best preserved late Jurassic sauropod tracks currently known.

In addition to sauropod foot casts and the abundance of interesting tracks at Dinosaur Lake, the Morrison Formation of the western states is also known for the (previously) controversial *Pteraichnus* track. *Pteraichnus* is a controversial track because it was first thought to be pterosaurian in origin, then it was attributed to a crocodilian, and now it is regarded as pterosaurian again. Since the original discovery of *Pteraichnus* in Arizona in the 1950s, similar tracks have been found in the Morrison Formation in Oklahoma, as well as in the aforementioned Summerville and Sundance formations. Pterosaurs are also known from skeletal remains in the Morrison, so the track and bone records are consistent in this respect.

Finally, we note one recent discovery of a series of parallel tracks of small to medium-sized bipeds from a site in southeast Utah, near the Arizona border. The tracks are very similar to tracks from the Late Jurassic and Early Cretaceous of Portugal and Spain that have been attributed to a type of ornithopod known as a

The Sauropod Straddle

Recent studies of the width or straddle of sauropod trackways show considerable variance from site to site and presumably from species to species. The first to take serious notice of this phenomenon was James Farlow, a dinosaur and tracks expert from Indiana University. Being a railway enthusiast, Farlow proposed in 1992 that the sauropod trackway widths or straddle could be characterized as "narrow-gauge" or "wide-gauge."

We agree with Farlow that distinct narrow- and wide-gauge trackways exist: the Morrison trackways at the Purgatoire site are narrow-gauge, while the famous Cretaceous sauropod tracks from Texas (chapter 5) are wide-gauge. Recent work suggests that all around the world narrow-gauge trackways are most common in the Jurassic, and wide-gauge trackways are most common in the Cretaceous. We have proposed calling the wide trackways *Brontopodus* ("brontosaur foot") and the narrow trackways *Parabrontopodus* ("toward *Brontopodus*").

It is intriguing to speculate that the straddle of sauropod trackways indicates diagnostic anatomical differences between different groups (families?) of sauropods. Consider that the width of living elephant and hippo trackways is very different and related to fundamental differences in anatomy (long legs and short body for elephants compared to short legs and long body for hippos). Although this example is not necessarily an ideal analog for sauropods, skeletal evidence confirms that sauropod anatomy did vary considerably, including leg length, body length, and pelvic width. According to sauropod expert Jack MacIntosh, wide-gauge trackways may be indicative of brachio-

(continued)

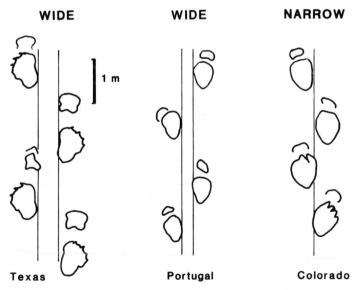

Examples of wide- and narrow-gauge sauropod trackways.

saurs—a tentative suggestion which Farlow considers reasonable, and with which we concur.

To date, most well-studied tracksites, including many that exhibit gregarious groups, reveal only one type of trackway gauge on any given surface. This appears to indicate that the trackway gauge is indeed an indicator of particular types (families?) of trackmaker (e.g., brachiosaurs, camarasaurs, diplodocids), that were either traveling in gregarious groups or frequenting a particular area at a specific time. We do not now know which taxonomic groups correspond to a particular straddle pattern, and we are not even sure that such a correspondence for extinct animals can ever be demonstrated with certainty. The prospects for making further headway with this line of inquiry are nevertheless intriguing and promising. Moreover, we suggest that an important principle may be at stake here. If straddle can help identify or distinguish sauropod groups, then it may be applicable to the study of trackways of other groups of vertebrates.

---

hypsilophodontid. The parallel trackways at the site in Utah and two sites in Europe provide compelling evidence that the trackmakers were gregarious. As many of the large Cretaceous ornithopods were gregarious (see chapter 5) it is perhaps not surprising that we are beginning to find evidence of herding among Late Jurassic ornithopod trackmakers.

Tracks Versus Bones

The Morrison Formation is the only Jurassic deposit so far discussed that contains a rich assortment of skeletal remains as well as a moderately good track record. What are the relative merits of the two types of evidence, body fossils and trace fossils?

There is a common myth prevalent in the popular (and even the technical) paleontological literature that skeletal remains and footprints rarely occur in the same geological units. Although there may be some truth in this, it is misleading as a universal generalization. The Morrison Formation provides a good example of a single deposit in which both bones and tracks occur, albeit rarely at the same sites. Generally, however, paleontologists attempting to reconstruct the populations and ecology of the late Jurassic ignore the Morrison track record and rely exclusively on the bone record for their conclusions, especially with regard to the overall composition and diversity of the fauna.

This practice is misleading for several reasons. First, at least for the Morrison dinosaur fauna, the total sample of skeletal evidence is probably quite severely biased toward large individuals and large species (i.e., if the Morrison community were anything like a normal modern animal community, there must have been many more small individuals and small species than the fossil record suggests). Second, dinosaurs and other vertebrates have been dug up from many different sites that represent different locations, different environments, different

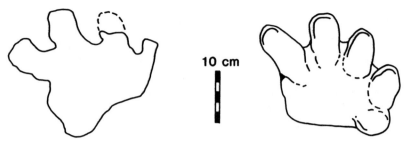

10 cm

FIGURE 4.56 Possibly the only known stegosaur track (left) is a front footprint named *Stegopodus* (Lockley and Hunt, 1998), from the Morrison Formation of Utah. Compare with hypothetical track proposed by Thulborn (1990).

times, and even different seasons. If we pool data from different sites, this can in theory mask differences that have biostratigraphical or paleoecological significance, and the practice can therefore inflate the estimated diversity of faunas. Such pooling also glosses over the relationship between the type of sediment (depositional environment) and the type of fossil evidence preserved. For example, many Morrison dinosaur bones and skeletons are preserved at the base of river channel deposits or in fossil soils (paleosols) where tracks are rarely found, whereas some of the best tracks occur along the shores of alkaline lakes or at the top of river channel sequences where bones are not usually found.

In the case of footprints, several sites within the Morrison do preserve relatively small tracks, perhaps providing a more accurate reflection of the proportions of small as well as large individuals and species. Also, as tracks are always in place and not transported, they may more accurately reflect the composition of local communities in different environments. For example, the virtual lack of stegosaur tracks suggests that stegosaurs were probably dryland or upland animals that stayed away from wetland areas where most tracks are preserved. Figure 4.56 shows a probable stegosaur track (*Stegopodus*) that we recently discovered near Moab, Utah. The suggestion that stegosaurs avoided wet ground is consistent with conclusions drawn from analyses of rather inconclusive data on the distribution of skeletal remains.

In the final analysis, all factors need to be considered, and tracks should not be dismissed just because skeletal evidence also happens to be available. It is perhaps understandable that paleontologists would dismiss track evidence if it were simply consistent with the bone record, thus telling us what we already know. But, in many cases, tracks provide additional ecological information that is not available from the bone record. Such information almost always has a bearing on questions that have to do with the completeness, or incompleteness, of the fossil record and the fact that what can be seen today is only a small and biased sample of the evidence that once existed.

Another recent discovery provides a further instructive example. Like stegosaurs, which are common in the Morrison Formation though their tracks are rare, crocodiles are also common, but their tracks, until recently, were unknown. A recent discovery in Utah reveals the presence of a large crocodilian described by John Foster and the senior author and named *Hatcherichus*. The name, in honor of John Bell Hatcher, was chosen because Hatcher reported a similar footprint from the Morrison Formation near Canon City, Colorado, in 1903, though he did not describe it in detail. The Utah specimen is part of a semi-walking, semi-resting trackway probably made when the animal was temporarily at rest at the bottom of a channel. Our studies indicate that this animal was larger than any known from skeletal remains in the Morrison Formation. So even in this case where crocodile skeletal remains are well-known, and we can predict that their tracks would be discovered, when we do find them, they indicate animals of unexpectedly large size.

The best of tracks and bones, even together, provide only the barest evidence of the biological richness and ecological complexities of past times. With so little to go on, we paleontologists should make use of every shred of evidence that can be gathered. Tracks surely will become an increasingly important part of that evidence. We will revisit this important theme as we continue our synthesis of track data in the next chapters.

Triceratops stomping grounds (based on tracks from the late
Cretaceous Laramie Formation, Denver area).

5

DAYS OF DINOSAUR DOMINANCE II: THE CRETACEOUS

*The country had never been worked over by any-
one with any sort of background in tracking down
dinosaurs.*

ROLAND T. BIRD, 1985,
IN *BONES FOR BARNUM BROWN*

■   The Cretaceous is the longest of the three Mesozoic periods. It lasted from 145 to 66 million years ago. Technically, the Cretaceous is divided into two epochs, Early (145–97 million years ago; corresponding to the Lower rock series) and Late (97–66 million years ago; corresponding to the Upper rock series). The period can be further subdivided into twelve geological ages, six in each epoch. Figure 5.1 shows these twelve ages in the form of the equivalent "stages" recorded in rock, with names of the most prominent formations that outcrop in and around Texas and Colorado.

Although the western United States has a nearly complete record of continental sedimentation and vertebrate life on land from Carboniferous time to the present, much of the early Cretaceous record is missing. Either conditions then were not suited for sedimentation or deposited sediments were later eroded away. In addition, a major marine incursion in the middle of the Cretaceous left a solely marine record in most rocks dating between about 97 and 77 million years of age. The Cretaceous track record in the western United States thus contains substantial gaps.

The large marine incursion into the western interior of the United States (one of the largest sea level rises of all time) produced what geologists call the Cretaceous Western Interior Seaway (figure 5.1). This seaway was quite different from the smaller-scale marine incursions during the Triassic and Jurassic (refer to figure 4.42). The Cretaceous incursion flooded into the continental interior from the Gulf of Mexico region in the south and from the Arctic region in the north—creating a continuous seaway from north to south persisted for at least 20 million years. A combination of sea level changes and earth movements caused the shorelines of the seaway to shift over rather large distances throughout this time. Thus, for example, the shoreline sometimes can be seen in the rock of the present-day state of Utah, sometimes in Colorado. These changes in shoreline position have proved relatively easy to document; geologists simply locate the transition from continental to marine deposits.

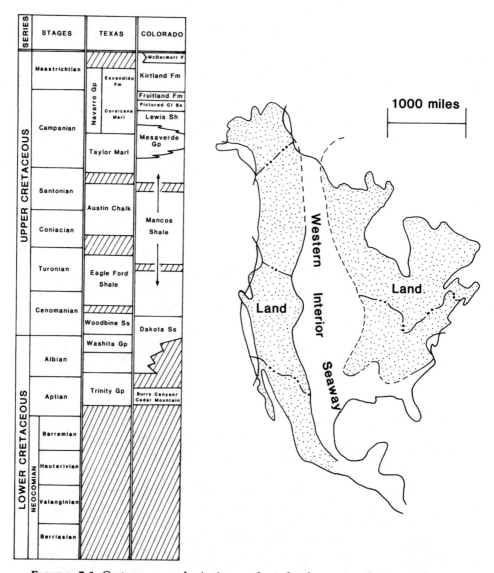

FIGURE 5.1 Cretaceous geologic time scale and paleogeographic map of North America. Stratigraphic columns, which include various track-bearing sequences, are shown for the general areas of Texas and Colorado. Hatched areas have no rock record.

Shorelines have an important bearing on the distribution of tracks because shoreline environments are an ideal setting for the accumulation of footprints. Moreover, the shoreline of the Western Interior Seaway, unlike the very localized lake, pond, and river settings that preserved tracks in other settings, was very extensive. Like the present day coastal plain of the Gulf of Mexico, this shoreline contained a mosaic of wetlands and waterways that provided suitable substrates for abundant tracks to be formed. And beyond the shoreline itself existed vast tracts of low-lying coastal plain that were also occasionally suitable for track preservation.

An Overview of Early Cretaceous Tracks

Figure 5.1 shows that a reliably identified rock record of the first four stages of the early Cretaceous is largely missing or poorly known from the western United States. We thus have to look elsewhere to get a sense of the kinds of tracks being made from the period of about 145–115 million years ago. Worldwide, early Cretaceous dinosaur tracks are dominated by large ornithopod tracks. These kinds of tracks were made by bipedal trackmakers, such as the famous *Iguanodon* and its relatives in the duck-billed dinosaur tribe. Many deposits in Europe, eastern Asia, and South America have produced such tracks from the earliest stages in the early Cretaceous, but few are known in North America.

Tracks can, however, be found in abundance in the western United States for the final two stages of the early Cretaceous—the Aptian and the Albian—which spanned 115–97 million years ago. Specifically, tracks are known from the Cedar Mountain Formation of Utah, the Glen Rose Formation of Texas (part of the Trinity Group), and the Dakota Group of Colorado, New Mexico, Oklahoma, Kansas, and Utah.

Without doubt, the Texas tracks are historically the most famous tracks of the early Cretaceous in the western states. The Texas tracks have given rise to controversial theories about dinosaur behavior, and new discoveries are still being made. The Dakota tracks have also proved to be very abundant. They are part of a huge complex of track-bearing coastal plain sediments that have been dubbed "the Dinosaur Freeway." (The Texas and Dakota tracks will each be discussed in separate sections, which follow.)

Less spectacular, but not to be overlooked, are tracks from the Cedar Mountain Formation, which have only recently been discovered. They are, nevertheless, already providing useful insights into the dinosaur communities that lived at an earlier stage in the early Cretaceous to the northwest of the famous Texas and Dakota sites.

To date, only three localities are known that yield tracks from the Cedar Mountain Formation. The first is a rather small tracksite near Moab, Utah, which contains an isolated theropod track and a few indistinct impressions. The second is found in association with the recently discovered, bone-rich Long Walk Quarry site, also in Utah, and has yielded several well-preserved iguanodontid tracks. The third site, also small, is near Thompson, just north of Arches National Park. This site has yielded several iguanodontid tracks, including three that were excavated and are now preserved in the collections at the College of Eastern Utah Prehistoric Museum (figure 5.2).

Although these three sites of the Cedar Mountain Formation have yet to be described in detail, they are important for several reasons. First, until recently, it has proved difficult to distinguish the Cretaceous Cedar Mountain Formation from the underlying Jurassic Morrison Formation, mainly because the strata in both formations are so similar. The discovery of bones and tracks has proved useful in demonstrating that each unit contains a very different fossil record. For example, in the case of the tracks, it appears that the Morrison contains very few examples of large ornithopod (iguanodontid) tracks of the type found in many early Cretaceous deposits. Thus, the distinction between the Morrison and Cedar Mountain track assemblages is significant, and it is a good example of the utility of tracks in biostratigraphy and paleoecology. Second, it is worth pointing out that all three Cedar Mountain tracksites were discovered during the course of prospecting and excavating skeletal remains. This observation underscores the fact that bones and tracks do occur together, as discussed in chapter 4.

In 1939 Sumner Anderson reported two sites with small, three-toed dinosaur tracks from the early Cretaceous Lakota Formation of the Black Hills area of South

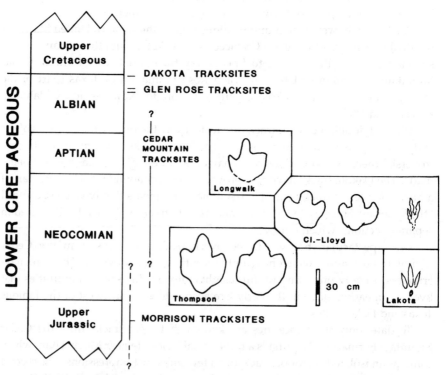

FIGURE 5.2 Tracks from the lower Cretaceous of eastern Utah (Thompson, Longwalk, and Cleveland Lloyd sites) and South Dakota (Lakota). These beds fill a large gap in the track record in western North America. They are mainly those of iguanodontids and a few small theropods.   *After Lockley and Hunt 1994.*

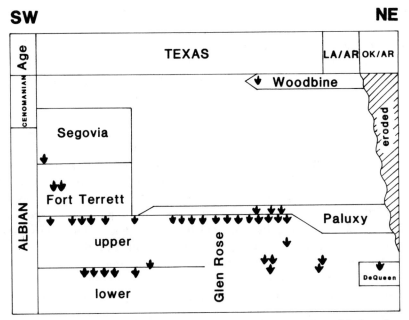

FIGURE 5.3 Cretaceous tracksites of Texas, with their stratigraphic location. *After Pittman 1989, 1992.*

Dakota. Most of the tracks reported then were small theropod footprints less than 15–20 centimeters long (figure 5.2). Although *Iguanodon* skeletal remains have been found in the Lakota Formation, iguanodontid tracks have not yet been found. The tracks in this area are thus in need of further exploration and study.

"Swimming" Brontosaurs and the Dangers of Misinterpretation

Most of the well-known dinosaur tracks of Texas are found in the Lower and Upper units of the Glen Rose Formation of middle to late Albian Age (about 115–105 million years ago). The Glen Rose Formation, part of the Trinity Group, outcrops over much of central Texas. The tracks tend to be distributed in discrete zones at the tops of particular strata within this formation (figure 5.3). The Glen Rose tracks are famous for the abundant sauropod and theropod footprints that have figured in various hypotheses, debates, and controversies about dinosaur social behavior, predatory relationships, and dinosaur swimming ability. Glen Rose tracks are also notorious as containing the putative "man-tracks" that creationists use as evidence for their belief that humans and dinosaurs walked the earth together—and only a few thousand years ago, as a strict reading of the Bible would have it. As we will discuss, many hypotheses as to dinosaur behavior have turned out to be fanciful. And the so-called man tracks are clearly the footprints of theropod dinosaurs. The Glen Rose Formation thus serves as a particularly useful illustration of how easy it is to misinterpret dinosaur tracks.

The best documented Texas tracksite, especially in terms of well-described footprint specimens, is the Paluxy River site, now managed as Dinosaur Valley

State Park. It is located near Glen Rose, southwest of Fort Worth. Here, in 1938, Roland T. Bird, assistant to Barnum Brown of the American Museum, first discovered and began investigating a dozen parallel brontosaur trackways (now named *Brontopodus*) and four theropod trackways, three of which followed the same trend as those of the brontosaurs (figure 5.4 shows a partial map). Subsequent excavation of the trackways of a single brontosaur and a single theropod (which were parallel or slightly converging) led to the acquisition of large display specimens by the American Museum (AMNH 3065) and the Texas Memorial Museum (TMM 40638-1), and the spawning of the hypothesis that the trackways indicate a single predator attacking its prey. This scenario was proposed partly because one of the theropod tracks in the sequence was reported as missing, possibly indicating that the predator had grappled with the brontosaur by latching onto the left rear flank, whence it was lifted off the substrate for a moment before resuming normal progression. Some trackers, including ourselves, consider this scenario dubious because of the regularity of the two trackways and the lack of evidence for significant changes in speed or direction that one might expect if a skirmish or attack had taken place. Moreover, the presence of evidence for so many other individuals passing through the area, at the same time seems to argue against a one-on-one attack. Based on the overall trackway picture, we surmise that twelve brontosaurs passed, probably as a herd, followed somewhat later by three theropods that may or may not have been stalking—but that certainly were not attacking.

The evidence for brontosaur herds, in general, seems good. A site at Davenport Ranch records at least twenty-three sauropods heading in one direction (figures 5.5 and 5.6). Such consistent trends are found at several Texas sites and probably indicate herds or groups perhaps progressing along shorelines, which would then result in preferred trackway orientations. Following the original suggestion made by Robert Bakker, many authors of popular texts have claimed that the Davenport Ranch tracks indicate a herd with large individuals (or adults) outside, protecting young or juveniles in the center. Careful analysis of the tracks by one of us (Lockley), the results of which were published in 1987, shows that an alternative hypothesis for the tracks is more likely. The trackways show that the twenty-three animals passed over a relatively narrow section of substrate. In order to do this they were not spread out on a broad front, but were rather walking in line with many of the smaller trackmakers following behind the larger individuals. Brontosaurs may well have exhibited parental care, but as yet the Texas tracks do not prove this.

Brontopodus from Texas is a wide-gauge sauropod trackway type. To the best of our knowledge, all the trackways from Davenport Ranch and from Dinosaur Valley State Park fall into this category. This is consistent with the hypothesis that herds of a particular species are represented at both sites. Moreover, it is possible, as suggested by James Farlow (who originated the "gauge" terminology to denote trackway widths of sauropods), that the trackmaker was a brachiosaurid. Brachiosaurids are distinguished from other sauropods by their long front limbs and relatively large front feet. The brachiosaurid *Pleurocoelus* is, in fact, known from skeletal evidence in Texas during the Albian stage of the early Cretaceous.

FIGURE 5.4 Map of parallel sauropod and theropod trackways along the Paluxy River at Dinosaur State Park, Texas. Several specimens were collected from this site: AMNH 3065 and Texas Memorial Museum specimen 40638–1. *After Roland Bird 1985.*

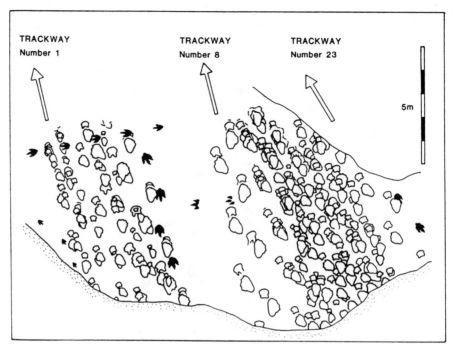

FIGURE 5.5 Parallel sauropod trackways, numbered 1–23, at the Davenport Ranch site, Texas. *After Roland Bird 1944.*

FIGURE 5.6 Roland T. Bird, discoverer of brontosaur tracks in the lower Cretaceous Glen Rose Formation of Texas, examining a tracksite at Davenport Ranch, where there is evidence for the passage of a herd. *Courtesy of the R. T. Bird collection.*

Trackways of brontosaurs showing mainly fore foot impressions—for example, footprints reported from the Mayan Ranch site near San Antonio (figure 5.7)—are puzzling and controversial. When first discovered, it was obvious that such tracks had not been made by an animal walking only on its fore feet. Roland Bird therefore proposed that the animal was swimming with its hind feet buoyed up by water. In part because of the prevailing notion, at that time, that brontosaurs were semi-aquatic swamp dwellers, the swimming brontosaur scenario was born. Through time this scenario has become deeply entrenched in the popular literature, and several subsequent interpretations of new trackways have followed Bird's interpretation, thus fostering the notion that trackway evidence for swimming dinosaurs is common.

FIGURE 5.7 A brontosaur trackway dominated by front foot (manus) impressions, Mayan Ranch site, Texas (after Bird 1941). Dotted outline of a single hind footprint represents a recently discovered underprint reported by Pittman (1989). Sketches show the competing interpretations (one swimming, the other walking) that have been proposed to account for the scarcity of hind tracks. *After Lockley 1991.*

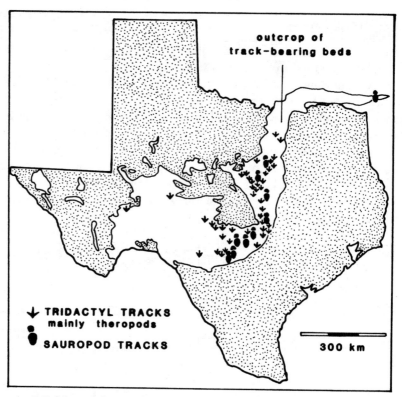

FIGURE 5.8 Map of the areal extent of dinosaur tracks in the Cretaceous (Albian) sediments of Texas (mainly the Glen Rose Formation and equivalents). *After Pittman 1989, 1992; Lockley 1991.*

But if the brontosaurs were not swimming, how does one account for the lack of hind prints? We believe the answer lies in the fact that the tracks are under-prints, not true footprints (see chapter 1 for a discussion of underprints). The original Mayan Ranch tracks were never collected, replicated, or measured and studied in detail probably because they were underprints rather than prints. They lack good definition. Recent work by Martin Lockley and Jeff Pittman have shown that the undertrack interpretation is correct and that, moreover, at least one hind footprint in the series can be discerned. All the evidence is, in fact, consistent with a walking rather than swimming interpretation and therefore discredits the swimming sauropod hypothesis. Notably, it has been shown that some sauropods carried a greater proportion of weight on their front feet than was previously supposed. And given that sauropod front feet were so small—about one quarter of the area of the hind feet—the front feet would indeed have made impressions or underprints in substrates that the hind feet might not have affected. As suggested to us by one of our colleagues, the weight carried by the front feet, and the animal's center of gravity, would vary, at least to some degree, as the long neck was extended forward or pulled back.

The swimming sauropod hypothesis has been further undermined, in our view, by recent discoveries of several brontosaur trackways in Europe that show only front footprint impressions. Because such trackways are now appearing more common we reason that they cannot all be explained by suggesting that an ever-growing number of sauropod trackway discoveries represent swimming animals. Finally, it is worth adding that studies of the anatomy and paleoecology of brontosaurs have also consistently shown that these sauropods were adapted to a terrestrial mode of life; they were not swamp dwellers. In fact, sauropod footprints have rarely been found in association with swamp deposits. Thus the notion that tracks indicate swimming sauropods can not be supported using any convincing line of available evidence.

We believe, moreover, that the swimming brontosaur example of false interpretation is instructive for wider purposes. The study of vertebrate tracks is a young and immature discipline. For this reason trackers should be very cautious about making sensational interpretations about the behavior of extinct animals based on track evidence that is merely suggestive. Otherwise they run the risk of discrediting the field. Evidence, legitimate interpretation, and wild speculation should be clearly distinguished, and, as in the case of the study of underprints, factors that control preservation of tracks in the geological record should be carefully considered before drawing conclusions about behavior.

Recent work by Jeff Pittman on the stratigraphy of the Texas tracksites has shown that about fifty sites are known, and these are mainly distributed at the top of the Upper Glen Rose Formation, with a lesser number at the top of the Lower Glen Rose Formation and only a few in other formations. These thin track-bearing layers extend over huge segments of ancient coastal plain. Two mega-tracksites that cover thousands of square miles are now recognized (figure 5.8). These track-rich, ancient coastal plain deposits include intertidal and lagoonal sediments closely associated with the Cretaceous Gulf—ancestral to the modern Gulf of Mexico. From Albian times onward, the Gulf was periodically enlarged as seas reached deeper into the continental interior, culminating in a Western Interior Seaway (figure 5.1). This seaway dominated the paleogeographical picture from Albian through latest Cretaceous times, when global sea level reached a record high.

Problematic Potholes

It is not often that dinosaur tracks are a hazard to the smooth operation of business in the modern world. However, a tracksite located in a gypsum quarry in southeastern Arkansas has produced irritating problems for the quarry operators. The tracks at this site, known as the Briar Site, were unrecognized until 1983, though heavy equipment operators had long found that certain surfaces were difficult to drive over. It turned out that the numerous potholes that were such a nuisance for quarry drivers were in fact dinosaur tracks.

The tracks were first recognized by Jeff Pittman. Pittman had been studying the sedimentology of the rocks at the Arkansas quarry when he happened to visit the famous Purgatoire tracksite of Colorado.

(continued)

There he saw examples of deep brontosaur tracks that resemble water-worn potholes in river valley bedrock. The challenge of trying to negotiate these deep brontosaur tracks is like no other driving experience. It is almost as if the surface had been designed for maximum irregularity, causing one's vehicle to lurch unpredictably this way and that. A much better view was obtained from Pittman's subsequent aerial survey which revealed ten parallel sauropod trackways, on an otherwise heavily trampled surface.

It became clear that two distinct surfaces in the Briar quarry each revealed thousands of tracks. These essentially trampled layers were (and are) continuously being exposed and destroyed by quarrying operations. In 1989, when Pittman began to recognize the existence of megatracksite zones in Texas, he demonstrated that the Arkansas tracksite occurs at the same level as the megatracksite complex at the top of the Lower Glen Rose Formation. Thus, although separated by a distance of more than 200 miles, the two areas are of the same geological age.

Tracks Along the Shores of the Western Interior Seaway

After the early Cretaceous seaway initially flooded the continental interior, subsequent pulses during the late Albian and early Cenomanian (98–97 million years ago) established the seaway as an broad oceanic strait connecting the Arctic and the Cretaceous Gulf (figure 5.1). The seaway's western shore ran north-south along the present Rocky Mountain trend; the eastern shore was in the mid-continent region. As the sea rose and fell, the shorelines expanded and contracted, resulting in regularly changing configurations of paleogeography.

One of the most extensive track-bearing shoreline deposit associated with the Western Interior Seaway is the Dakota Group. Its uppermost layers comprise another megatracksite complex of coastal plain deposits, including sedimentary facies attributable to rivers and lakes as well as to marine shorelines. Detailed studies have shown that dinosaur tracks occur in these uppermost layers from northern Colorado, near the town of Boulder, to eastern New Mexico, almost as far south as the town of Tucumcari (figure 5.9). At Clayton Lake State Park, in northeastern New Mexico, one of the larger exposed tracksites is accessible to dinosaur enthusiasts and other visitors. Another protected site, at Dinosaur Ridge west of Denver (figure 5.10), has been designated a National Natural Landmark. This tracksite is very near the historic Morrison Formation bone quarries that yielded *Apatosaurus* and *Stegosaurus* in 1877. The juxtaposition of bone and tracksites makes this an important paleontological area, and interpretive signs, brochures, and *A Field Guide to Dinosaur Ridge* are available to help visitors understand the dinosaur resources in the area.

There are so many tracksites known from the Dakota Group of the Colorado Front Range and southern high plains that it is impossible to describe them all in the space available. And new discoveries are being made all the time. Rather, we will concentrate on just a few representative tracksites, trackways, and tracks.

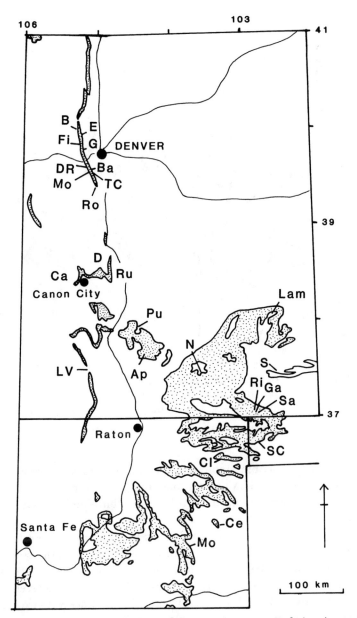

FIGURE 5.9 The distribution of Dakota Group dinosaur tracksites in portions of Colorado, New Mexico, and Oklahoma. Stippled area represents outcrops of the Dakota Group. Letter abbreviations refer to geographical names of thirty tracksites, including the eight that are discussed and illustrated in this chapter: Eldorado Springs (E), Golden (G), Dinosaur Ridge (DR), Turkey Creek (TR), Roxborough (Ro), Lamar (Lam) Clayton Lake (Cl) and Mosquero Creek (Mo). *After Lockley et al. 1992.*

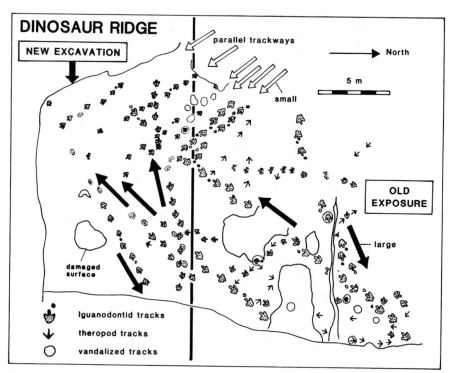

Figure 5.10 Map of the Dinosaur Ridge tracksite, Dakota Group, near Denver, Colorado, showing where overburden was removed in 1993 from the south (left) portion of the area, exposing many new trackways. *After Lockley and Hunt 1994.*

We will begin by describing representative northern tracksites near Denver and then proceed southward. Although most tracks are attributable to dinosaurs, some sites contain a few footprints attributable to birds and crocodilians, and we will mention these too.

We began our studies of Dakota tracks by examining about ten dinosaur tracksites along a 65 km stretch of outcrops in the eastern foothills of the Colorado Front Range between Boulder and Roxborough (figure 5.11). We found that most of the track-bearing layers occur at about the same level in the rock sequence. We also collected several dozen tracks and trackways excavated from these sites (both originals and replicas) and have preserved them in the University of Colorado at Denver—Museum of Western Colorado collections.

The largest site from this northern series of outcrops that we have yet examined is a commercial clay quarry in the vicinity of Eldorado Springs. Here, multiple track-bearing layers are preserved, including a few extensive surfaces with tracks of ornithopod and theropod dinosaurs. Although it is not practicable to map all of the highly trampled surfaces, a few localized areas have yielded clearly defined tracks, from which the map in figure 5.12 was prepared. When traced southward, similar clay quarries also yield ornithopod and theropod tracks of the same type.

A clay quarry near Golden is particularly noteworthy as the first site to have ever yielded bird tracks of Mesozoic age. (Bird tracks are distinguishable from dinosaur tracks in that they are generally smaller, with widely splayed and very slender digit impressions.) These bird tracks were first reported in 1931 by Maurice Mehl, and they were named *Ignotornis*. Subsequently, collectors removed many specimens and the locality was reported to be barren—its exact location forgotten. However, in 1988 we rediscovered a bird track locality where

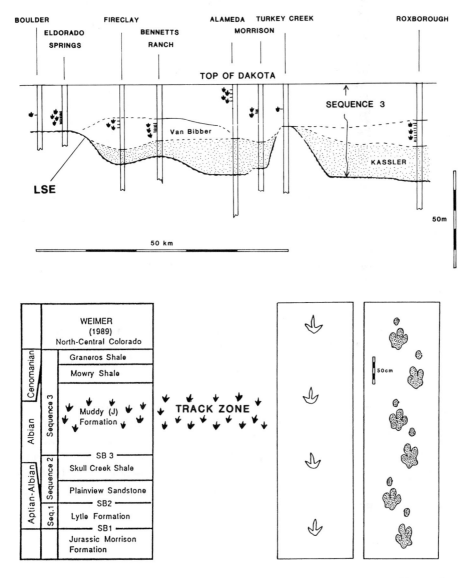

FIGURE 5.11 Track-bearing levels in the Dakota Group between Boulder and Roxborough, Colorado. Stippled area represents channel deposits without tracks. *Bottom left:* Relationship of track-bearing zone to Dakota sequences. *Bottom right:* Detail of common track types, a slender-toed, coelurosaur-like theropod trackway and a broader-toed trackway of a quadrupedal ornithopod (*Caririchnium*). *After Lockley et al. 1992.*

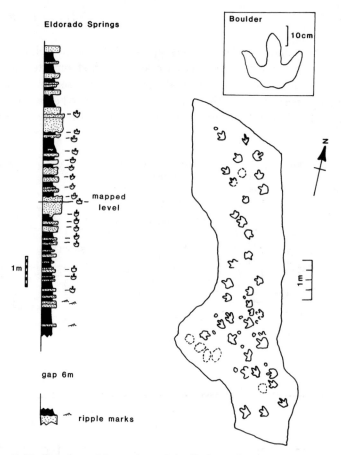

Figure 5.12 Stratigraphic section of the Dakota Group near Eldorado Springs, Colorado, with track-bearing layers marked. Map of one level is shown, with inset of detail of track cast recovered nearby.
After Lockley et al. 1992.

dinosaur tracks also occurred (figure 5.13). Coincidentally, a specimen from this site also proved to be the first in the world in which bird and dinosaur footprints were recorded on the same surface. This unique specimen, a slab weighing about three metric tons, was excavated in 1989 by our research group, in conjunction with a paleontologist from Japan, as part of a cooperative research and exhibits project. The specimen then went on a tour of various science museums in Japan as part of an international "tracking dinosaurs" exhibit.

Parallel Trackways and Dinosaur Herds

We have already mentioned and illustrated the Dinosaur Ridge site (figure 5.10), situated alongside Alameda Parkway west of Denver, and discovered when the road was constructed in 1937 (figure 5.14). As shown by the map, this site (like the ones to the north) is characterized by an assemblage of theropod and ornitho-pod tracks. The trackways of the two largest ornithopods at this site are parallel. At least three trackways of medium-sized individuals parallel one another in a different direction. Such trackway patterns provide evidence of gregarious behavior among the trackmakers. Our detailed study of these ornithopod trackways indicates that they are different from most typical iguanodontid tracks because they were made by trackmakers that consistently walked on all fours. This contrasts to the traditional view of *Iguanodon* as a bipedal animal. For this reason we have assigned the Dakota tracks to the ichnogenus *Caririchnium*, rather than to the category of *Iguanodon* tracks.

Again, we find that tracks at Dinosaur Ridge occur at multiple levels (about six) and on bedding planes that are exposed for several hundred yards along the roadside (figures 5.11 and 5.15). As is the case at many such sites, multiple track levels indicate that dinosaurs repeatedly returned to this coastal plain habitat and that conditions for track formation and preservation were favorable for prolonged periods. This particular area, a National Natural Landmark and excellent outdoor teaching laboratory for students, has been intensively studied over the years. Every layer of strata has been given a numerical or alphabetical designation and invertebrate trace fossil experts have identified as many as twenty different types of burrows and trails attributable to invertebrates. But it is surprising that tracks at most of the six levels were not recognized until recently.

A few miles south of Alameda, at a site known as Turkey Creek, we have documented tracks at other levels of the Dakota Group. Our most significant discovery at this new site has been tracks of ornithopod dinosaurs that display skin impressions (figure 5.17). Such tracks are very rare in Cretaceous deposits, and the Turkey Creek example is one of only two so far documented in western North America.

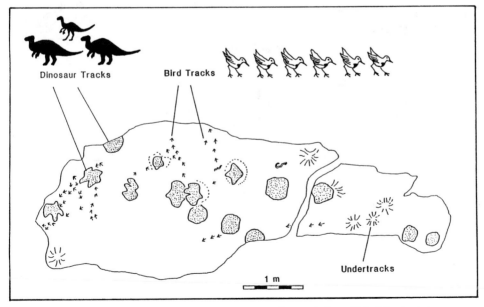

FIGURE 5.13 Dinosaur tracks (with some undertracks) and bird footprints occur on the same surface in the Dakota Group at a location near Golden, Colorado.   *After Lockley et al. 1989, 1992.*

FIGURE 5.14 Historic photograph of first tracks uncovered in 1937 from the area now designated as Dinosaur Ridge (Cretaceous, Colorado). Although originally interpreted as a theropod trackway, it is probably an ornithopod trackway. Note that the trackway was painted to improve it for the photograph. *Courtesy of the Denver Museum of Natural History.*

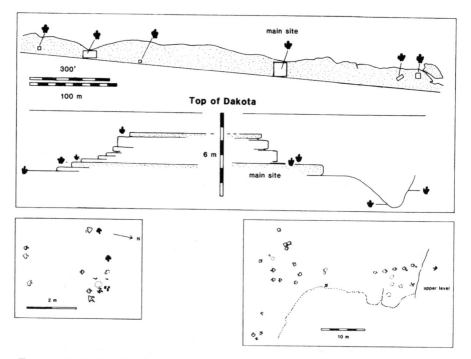

FIGURE 5.15 Distribution of track-bearing levels in the Dakota Group at Dinosaur Ridge, near Denver, with detail of two tracksites from the south (left) end of the outcrop. *After Lockley et al. 1992.*

Digging for Dinosaur Tracks

Despite the fact that there is a time-honored paleontological tradition of "digging for dinosaurs," surprisingly few large-scale excavations of dinosaur tracks have ever been undertaken. Unlike dinosaur skeletons, which have to be excavated and restored, trackways are typically found on the surface, so no digging is required, and usually they are then mapped, measured, and left in place. In fact, in most cases, it is destructive to the integrity of a site to remove individual tracks, as can be seen at Dinosaur Ridge and other sites that have suffered vandalism. Anyone entertaining thoughts of removing a large tracksite for safekeeping or study soon realizes that such plans are prohibitively expensive and logistically very difficult. For this reason there have been very few track excavations beyond the fortuitous discovery of tracks during various quarrying operations. It is, however, feasible (though still expensive) to remove layers of overburden from above known track beds and thus to expose enlarged track-bearing surfaces. Even so, rarely have tracks been considered sufficiently important to do this on a large scale.

In 1940 Roland Bird was the first to undertake a large-scale track excavation. He removed segments of the famous sauropod and theropod trackways from the bed of the Paluxy River, now protected in Dinosaur Valley State Park in Texas. He then shipped the slabs to the American Museum of Natural History in New York City. This Paluxy River excavation was a massive undertaking since it involved removing the track bed, not just the overburden. As Bird recalls in his autobiography, months of labor were involved in chipping around the selected area, and then removing it in hundreds of blocks that would later be painstakingly pieced back together. Bird even worried that his precious trackways might be torpedoed by a German U boat as they were shipped along the coast from Houston to New York.

Until recently, the only other large-scale track excavations involved just the removal of some light overburden at Dinosaur State Park in Connecticut during the development of the site, and removal of overburden at a relatively small site called Lark Quarry in Australia. As part of the ongoing efforts to develop Dinosaur Ridge as an outdoor interpretive center, our trackers research group of the University of Colorado at Denver, in conjunction with the Friends of Dinosaur Ridge and with the support of the Jefferson County Scientific and Cultural Facilities District, organized an excavation to remove overburden from the area just south of the main tracksite (figure 5.15) and thus to expose fresh tracks. Beginning in mid November 1992, a one-meter-thick layer of sandstone above the main track surface was removed, using precision blasting, from an area measuring about 200 square meters. Beneath this sandstone is a layer of shale about thirty centimeters thick. This shale layer covers the main track layer impressed in a layer of sandstone (or "target layer") and thus protected it from damage during the blasting and removal of thousands of tons of rock.

(continued)

We nevertheless anticipated that the shale itself might contain tracks. Our hopes were amply rewarded by the discovery of natural sandstone casts preserved in the base of the overlying sandstone layer. Although the sandstone was being removed as rubble, the precision blasting was done with sufficient care to allow some sandstone casts to remain in the top of the shale as the overburden was removed. We then mapped these casts and collected them, adding a new layer of information about the surface above the target layer (figure 5.16).

At the time of writing, the shale has been cleaned and we have finished a first draft of a map of the new area (figure 5.10). At first it was hard to see any but the large ornithopod tracks, but after a single winter exposed to Colorado weather, the site began to show more of the smaller tracks of theropods and ornithopods. From this experience we gain the opportunity to see how tracks weather through time and have learned that it is very helpful to monitor and revisit sites in order to acquire the maximum amount of available information.

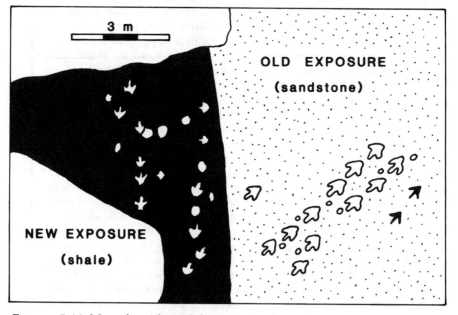

FIGURE 5.16 Map of ornithopod dinosaur tracks revealed by the 1992–1993 excavation at Dinosaur Ridge.

Further south, at Roxborough State Park another assemblage of tracks has been documented from a sequence of multiple track-bearing levels (figure 5.18). When first discovered in the mid 1980s, the tracksite locality was outside the park, but through fortunate circumstances the park was able to obtain the land and include the site within its boundaries.

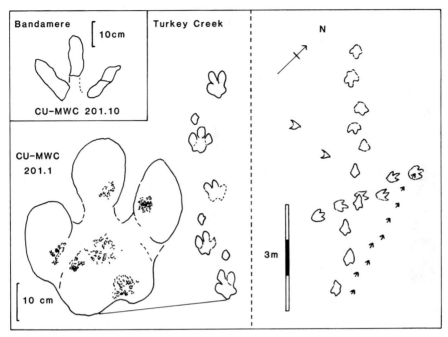

FIGURE 5.17 Ornithopod dinosaur tracks with skin impressions, and a map of trackways from the Dakota Group at Turkey Creek, near Denver. Inset shows a slender-toed theropod track from a nearby site. *After Lockley et al. 1992.*

Our research to date has shown that the Dakota Group of the eastern foothills of the Colorado Front Range—especially between Roxborough and Boulder—is replete with tracks occurring at multiple levels in the upper part of the Dakota Group (figure 5.11). These same levels also reveal tracks at localities stretching far to the south, for example, near Colorado Springs and across the southern high plains in the vicinity of Lamar, Springfield, and Campo in southeastern Colorado. Tracks excavated in the 1930s from the Lamar site (figure 5.19) are now on display at the Denver Museum of Natural History.

These same track-bearing strata extend into northeastern New Mexico; the track beds of Clayton Lake State Park also occur at about the same level, near the top of the Dakota Group. At this site, which is accessible to the public, paleontologists from New Mexico have documented more than 500 tracks. Farther south is another large site, named Mosquero Creek, which was discovered by local cowboys. Here we ourselves have discovered additional compelling evidence of gregarious dinosaur behavior in the form of at least 55 parallel northward-trending trackways of small quadrupedal ornithopods (figures 5.20 and 5.21). This site also contains ten trackways of larger, bipedal ornithopods that were heading in the opposite direction. This evidence is reminiscent of shoreline trackway patterns, in which animals often travel in two opposing directions. The evidence, like that just described from Dinosaur Ridge, also indicates that small and large animals (presumably representing different age groups of the same species because the tracks are identical except for size) often traveled in separate groups.

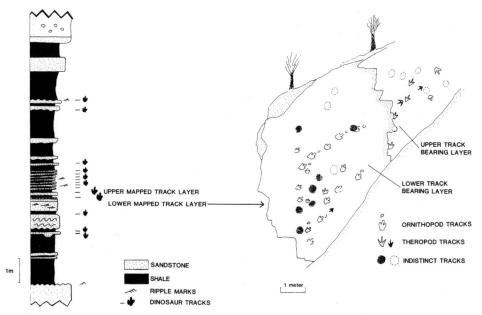

FIGURE 5.18 Stratigraphic section of a part of the Dakota Group at Roxborough State Park, Colorado, plus a map of two adjacent track-bearing levels. *After Lockley et al. 1992.*

Overall, the Mosquero Creek site is exceptional because it reveals such a large number of parallel trackways. Some of the aforementioned Mesozoic sites, like Davenport Ranch and the Paluxy and Purgatoire localities, also each provide evidence of at least a dozen parallel or subparallel trackways, but rarely are the exposures large enough to indicate whether or not the trackways formed part of a much larger group. At the Mosquero Creek site, although it is a relatively small area, we have mapped the largest concentration of parallel trackways of large ornithopods ever recorded from a single surface. Moreover, the pattern of the outcrop of track-bearing surfaces suggests that there are many adjacent areas where additional trackways eroded away. Thus, the total of 55 recorded, parallel trackways is clearly a minimum estimate. Not all the trackways are perfectly parallel, but most conform to two slightly divergent trends. This suggests that there may have been two waves, or groups, of trackmakers that went by at slightly different times. We measured the depths of footprints in all the trackways and obtained very consistent measurements, which would suggest that the animals had all passed at more or less the same time (that is, when the substrate had the same consistency). If this interpretation is correct, then perhaps the waves of trackmakers suggest a large herd in which smaller groups passed by veering in slightly different directions. Certainly the animals did not all walk on a broad front, but instead were evidently strung out fore to aft to some degree.

The large southward-trending trackways are deeper than the small northward-trending trackways and reveal no manus impressions. For this reason the two sets could be attributed to different species. However, such an interpretation may not necessarily be correct, for several reasons. First, it is possible that the front footprints were overprinted by the larger hind tracks. Second, it has been

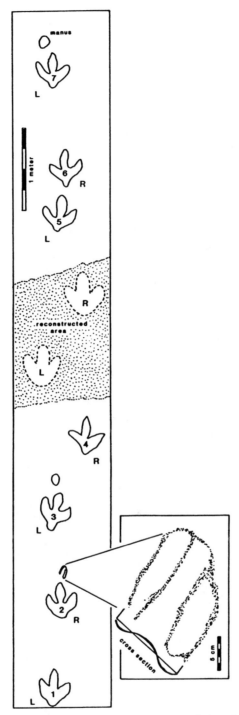

FIGURE 5.19 Tracks from the Dakota Group near Lamar, Colorado, are on display at the Denver Museum of Natural History. Note detail of front footprint (no. 2). The Museum reconstructed two missing tracks and mounted an ornithopod skeleton on the tracks, as if walking along. *After Lockley 1987.*

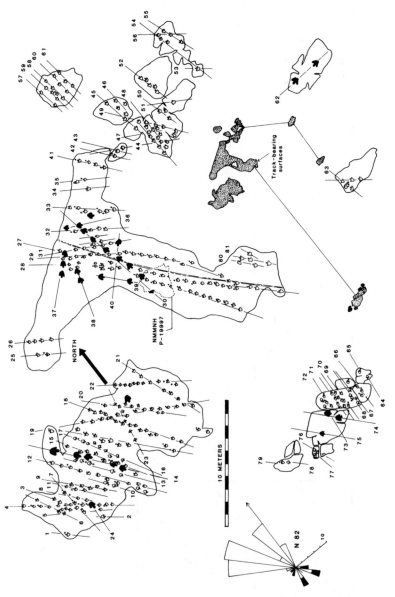

Mosquero Creek Tracksite

FIGURE 5.20 Dinosaur trackways from the Dakota Group at Mosquero Creek, New Mexico, reveal evidence of gregarious behavior among ornithopods. Map shows at least 55 parallel to subparallel trackways of small quadrupedal individuals trending NW. Trackways of larger bipedal individuals (black) all trend into the southern sector. Trackway numbers (1–81) are located at the distal end of trackways, i.e., indicating the direction in which the trackmakers were heading.

suggested that ornithopods were "facultative" or optional bipeds. They could walk either on their hind feet or on all fours as they chose, or they could switch back and forth. The trackway from the Lamar area of Colorado (figure 5.19) appears to show an alternating bipedal/quadrupedal pattern, although the presence or absence of front footprints in such examples can also result from differences in substrate consistency and preservation.

One of the small ornithopod trackways at the Mosquero Creek site shows a regular alternation of long and short steps. The placement of the animal's front footprint relative to its hind foot is different in alternate steps (figure 5.22). This is likely an example of a limping dinosaur. By coincidence, before this limping pattern was recognized, the New Mexico Museum of Natural History in Albuquerque had removed a section of the surface that included a short section (part of four consecutive manus-pes sets) of the limper's trackway. Thus, an example of the limping trackway is now easily accessible for the public to see.

Overall, the Dakota Group, from outcrops in Boulder and Eldorado Springs in the north to Mosquero Creek in the south, proves to be a track-rich zone that

FIGURE 5.21 Two parallel trackways at the Mosquero Creek tracksite. The trackway on the right indicates a limping animal.

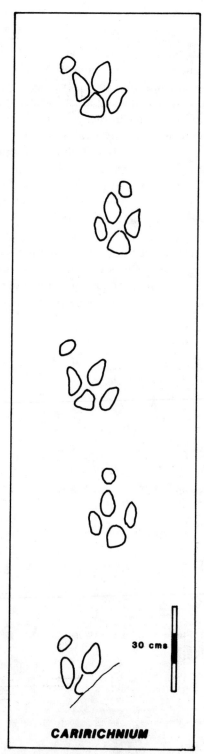

FIGURE 5.22 Drawing of the Mosquero Creek ornithopod trackway (figure 5.21) that shows a limp. Note the alternating long and short steps.

Mama, Papa, and Baby Dinosaurs

How can ichnologists determine whether small and large dinosaur tracks of a particular type represent different species, juveniles and adults of the same species, or perhaps different sexes? (Some dinosaur experts suggest that males were larger, as in many modern mammals; others argue that the females were bigger, as in many frogs.) We touched on this subject in chapter 4, where we noted that some trackers have claimed that *Grallator* and *Eubrontes* tracks simply represent small and large individuals of the same species. We disagreed with that interpretation, largely on the basis of the different shape of the tracks, not to mention the fact that these two track types have been classified separately for over a century.

In the case of the multiple iguanodontid trackways (*Caririchnium*) that we discovered and named in the Dakota Group, there are no obvious differences in shape between the large and the small tracks. In addition, the tracks all come from a very narrow zone of strata, inferred to represent a short period of time. At some sites, for example Dinosaur Ridge and Mosquero Creek, small and large tracks of *Caririchnium* occur on the same surfaces, in arrangements that indicate herd activity.

In a recent study of Cretaceous duck-billed dinosaurs—close relatives of the ornithopods responsible for making the Dakota tracks—Jack Horner of the Museum of the Rockies (Bozeman, Montana) has estimated the age of ornithopods from the size of their skeletons. This has led us to ask the next logical question. If Horner can estimate dinosaur age from the size of skeletons, might we also be able to do it from footprints? In theory, the answer is yes. The graph portion of figure 5.23 is Jack Horner's calculation of the estimated age that corresponds to a given size of duck-billed dinosaur. The figure shows that ornithopod dinosaurs grew very quickly and were probably close to half their full size by their first or second birthday. Using this graph, and assuming that the biggest track was made by an adult, we can make some age estimates for *Caririchnium* tracks. We conclude that the smallest tracks (about 20–25 cm long, or half full size) were made by dinosaurs that were only about a year old (figure 5.24). Thus, according to the Horner model, the smaller Mosquero Creek trackmakers may have been yearlings; the larger animals, would have been adults, at least 5–7 years old. We do not claim that these age estimates are absolutely precise, but they are as accurate as we can get at present.

In the future, as estimates of dinosaur age improve, tracks will surely become useful in estimating age distributions of populations. This is because tracks often occur in large numbers, representing herds, and are easy to measure accurately and quickly. Studies of tracks elsewhere in the world have indeed shown that a lot more about the distribution of large (adult) and small (juvenile or baby) dinosaurs can be learned from tracks than from bones. For example, in Dakota-aged deposits in South Korea, and at other sites around the world, baby brontosaur tracks are much more common than baby brontosaur skeletons.

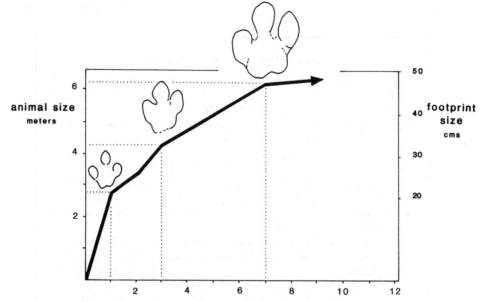

FIGURE 5.23 Size range of ornithopod tracks from the Dakota Group of Colorado, Oklahoma, and New Mexico, plotted against the dinosaur growth curve developed by Jack Horner. This diagram may allow us to estimate the age of trackmakers from their footprints.

FIGURE 5.24 Ornithopod track cast (*Caririchnium*) from the Dakota Group of eastern Colorado indicates a small animal that was barely half of full adult size.

extends for hundreds of miles along the trend of the western shoreline of the Western Interior Seaway. This means that the track-bearing beds of the Dakota form another coastal plain megatracksite system whose full extent remains to be explored.

The Dinosaur Freeway

The concept of the Dinosaur Freeway evolved as the extent and configuration of the Dakota megatracksite became apparent. From Boulder and Denver in the north to Clayton and Mosquero Creek in the south, most of the dinosaur tracks are very similar. They are predominantly those of *Iguanodon*-like ornithopods. Many are virtually identical to the footprints of an early Cretaceous ornithopod from the Carir Basin of Brazil and have therefore been named *Caririchnium* (track from Carir; see figure 5.11). Others are slightly or subtly different, resembling so-called *Iguanodon* tracks from early Cretaceous strata in England and other parts of the world. Throughout the Dakota megatracksite we also find the tracks of carnivorous dinosaurs. Most of the well-preserved examples are those of medium-sized, slender-toed, gracile theropods of the bird-like type (figure 5.11).

The fact that most tracksites in the Dakota Group occur at about the same stratigraphic levels (figure 5.11) and reveal the same types of tracks suggests a geographically widespread, ornithopod-dominated dinosaur community occupying this particular coastal plain habitat for an extended time. (It must have taken many thousands, even millions, of years for the multiple track-bearing levels to accumulate). Bearing in mind the fact that dinosaur bones are unknown from any of these sites, tracks are extremely important in demonstrating the prolonged existence of ornithopod communities in this region at this time. This conclusion fits well with the observation of John Ostrom of Yale University, who inferred that Cretaceous ornithopods preferred humid, well-vegetated coastal plain environments. Iguanodontid dinosaurs are an early representative of the duck-billed type of dinosaurs, and, as we shall see, track evidence shows that their descendants, the true late Cretaceous duck-billed dinosaurs (hadrosaurs), continued to inhabit these lush and humid habitats.

Our study of the many Dakota tracksites began in 1987 at the Alameda Parkway site—now known as Dinosaur Ridge. Because Alameda means "promenade" in Spanish, the idea of a dinosaur promenade evolved. When a new section of north-south freeway (Colorado 470) was constructed parallel to the parkway and we began to demonstrate the extensiveness of the north-south belt of tracksites, it was inevitable that this Cretaceous megatracksite would be dubbed the Dinosaur Freeway (or Cretaceous 470). Several authors have speculated that this coastal plain thoroughfare was once a dinosaur migration route. This is a distinct possibility, but hard to prove. We know only that the same dinosaur types (community) frequented the whole area. How far individuals or local populations ranged is still a matter of speculation.

One final observation is pertinent to our explanation of where tracks occur in the Dakota Group. Because the Dakota Group has produced large amounts of oil and gas, it has been extensively studied by several generations of exploration geologists and academic stratigraphers and sedimentologists. This geological research has shown that the Dakota Group can be divided into three widely recognized stratigraphic sequences (figure 5.11) that can be recognized throughout the megatracksite area. Our studies have shown that all the tracks occur in the upper sequence (number 3, shown in figure 5.11). Although we do not fully understand the reasons for this, the phenomenon is analogous to the situation reported, in chapter 3, for the Chinle Group, where almost all the tracks occur in the upper sequence. There may also prove to be other analogs, such as the occurrence of the Entrada-Summerville and Glen Rose megatracksites within specific zones that are associated with stratigraphic boundaries.

At present the most obvious connection that we have noted in the distribution of tracks between the Chinle and the Dakota Group is that in both cases the upper, track-rich levels are associated with what geologists call aggradational deposits (i.e., deposits that aggrade or build up in low-lying areas to form a regular grade or plain, usually at or near the watertable or what is called "base level"). The same is probably true for the Moab megatracksite at the Entrada-Summerville transition; like the Dakota track beds, it represents an aggradational deposit associated with the onset of a rise in sea level. This indicates that the preservation of tracks in the geological record is controlled by sedimentological processes such as aggradation, which is in turn generally associated with rising watertables or sea level (= rising base level). Although biological and ecological factors also come into play in determining which habitats are preferred by which dinosaurs, we must not forget that the presence or absence of tracks and other fossils may be determined to a significant degree by sedimentological processes and accumulation patterns.

Having now introduced three megatracksites, and having demonstrated that each contains a distinctive assemblage of dinosaur tracks, it is worth returning to the "ichnofacies concept" introduced in earlier chapters. Every time multiple examples of sites with the same types of tracks repeatedly occur in the same type of sedimentary rock, there is, *by definition*, an "ichnofacies," and it is proper to explore what such patterns mean. The simple answer is that the same types of animals were active, repeatedly, in a particular type of environment. According to this outlook, the same ancient animal communities from a particular and quite distinctive ecosystem or ancient environment are represented in each case. The similarities in fossils are not just a random occurrence at unrelated tracksites. The *Brontopodus*-bearing limestones of Texas and the *Caririchnium*-dominated sandstones and shales of eastern Colorado, eastern Oklahoma, and northeastern New Mexico provide two fine examples.

In a 1993 paper we defined the *Brontopodus* ichnofacies, based on the Texas sites, and noted that all around the world similar Jurassic and Cretaceous limestones often contain concentrations of *Brontopodus* or *Brontopodus*-like tracks. This does not mean that every brontosaur tracksite is part of the *Brontopodus* ichnofacies. However, it does mean that multiple examples of limestones with brontosaur tracks, such as are found in late Jurassic deposits in Portugal and Switzerland, should be examined for similarities. All these deposits qualify as examples of this ichnofacies; together they suggest that brontosaurs frequented the same types of environment for much of the late Mesozoic. Specifically, these environments were low-latitude settings characterized by dry to semi-arid seasonal climates.

By contrast, brontosaur tracks are distinctly sparse or absent in sediments that represent humid, well-vegetated, mid-latitude or high latitude environments. Instead, tracks of large ornithopods are abundant in those sediments. This pattern is so robust that in the case of the Dakota we find *Caririchnium* at almost every one of the thirty known sites, but we have yet to find a single brontosaur track in the Dakota. Thus the Dakota megatracksite has been named the *Caririchnium* ichnofacies. It is typical of many more ornithopod ichnofacies found in similar environments around the world.

One of the interesting benefits of becoming familiar with vertebrate ichnofacies is that we can begin to predict what we will find in a particular sedimentary rock unit. This will alert us to look closely at anything that is odd or anomalous. For example, what would we think if someone reported a site with eight or ten different dinosaur track types from the Dakota Group in the middle of the Dinosaur Freeway, where everywhere else there are normally no more than two track types? Either the report would signify an unusual local pocket of high dinosaur diversity or it would suggest that someone had been overzealous in interpreting different track types.

Overzealousness is exactly what has happened, in our opinion, at Clayton Lake State Park in New Mexico. At this site is a beautifully designed outdoor interpretive trail for the benefit of visitors. But the interpretation is exaggerated and refers to a diversity of eight different track types, including tracks of a web-footed theropod (figure 5.25). This would be fascinating if it proved to be a defensible interpretation, but in reality the trackway is simply that of an ornithopod (*Caririchnium*) made on a relatively soft substrate. In fact much of the site is trampled, causing many tracks to appear to have odd shapes. Our observations suggest that only two dinosaur track types are at this site; the signs erected for the public tell a rather farfetched story.

Early Cretaceous Bird and Crocodile Tracks

Until recently it has been assumed that bird tracks were rare in rocks formed before the Cenozoic. Indeed, not a single bird track was known from the Mesozoic until M. G. Mehl discovered the very first example in the Dakota Group near Golden in 1931. He named these tracks *Ignotornis* (figures 5.26 and 5.27) and attributed them to a small shorebird or wader.

Although only one other authenticated bird tracksite (also *Ignotornis*) has been discovered in the Dakota Group, several dozen bird tracksites have been found in other Cretaceous formations—including many in eastern Asia and a few in western North America. After the *Ignotornis* specimens, the best-documented North American examples are reported from the Gething Formation of western Canada. The Gething deposits are slightly older than the Dakota Group. Together, these glimpses of early Cretaceous avian life suggest that birds were more established and diversified than could be deduced from the sparse record

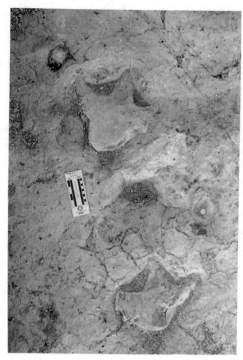

FIGURE 5.25 Seemingly web-footed theropod tracks from Clayton Lake State Park are in fact trackways of quadrupedal ornithopods (*Caririchnium*), typical of the Dinosaur Freeway area. Note manus imprints with 25-cent coins in center.

of avian skeletal remains known from anywhere in the world at this time. As we shall discuss, bird tracks are now also known from several later Cretaceous deposits.

In 1997 Yuong-Nam Lee reported a very large (19–21 cm long) slender-toed bird track which he named *Magnoavipes,* from the Cenomanian, Woodbine Formation of Texas. In our opinion these are the same track type known from certain Dakota Group sites in Colorado. Thus large waterbirds may have been present at the same time as the *Ignotornis* trackmaker.

Overall, the Dakota Group occupies a special place in the history of research into dinosaur and bird tracks. However, these kinds of tracks are not the only footprints reported from Dakota Group strata. Crocodile (or crocodile-like) tracks are now known from at least four localities in Colorado, Utah, New Mexico, and Kansas. But some interpretations have proved controversial. The Colorado tracks were first interpreted by some authors as the foot marks of aigialosaurs (marine lizards related to mosasaurs); the New Mexico tracks were once thought to be pterosaurs; the Kansas tracks were believed to be the traces of swimming ornithischian dinosaurs.

The case history of discoveries and interpretations of the New Mexico tracks at Clayton Lake is particularly interesting. First they were considered by David Gillette and David Thomas as the trail of a pterosaur taking off because they appeared to show an irregularly spaced arc of three-toed tracks. This same pattern was later attributed by Lockley to a swimming crocodile (figure 5.28). Finally, as more tracks emerged through erosion and weathering, the tracks were reinterpreted by Chris Bennett as part of an assemblage of crocodilian trackways. As discussed in chapter 4 in the case of Jurassic *Pteraichnus* tracks, this is not the first time that controversy has arisen over the identity of purported pterosaur tracks. In this case, however, the crocodile interpretation appears to be correct.

Overview of the Late Cretaceous Environment

Between the Cenomanian stage that marks the beginning of Upper Cretaceous rock strata (the late Cretaceous, 97 million years ago) and the Campanian stage (that began 84 million years ago), much of western North America was submerged beneath the Western Interior Seaway. The deposits that accumulated

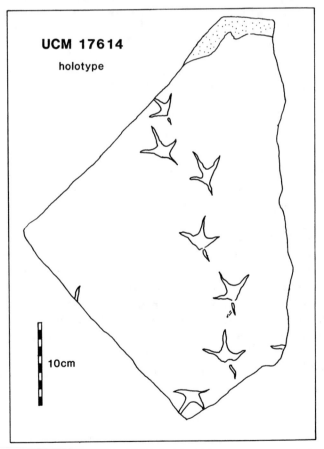

FIGURE 5.26 Mesozoic bird tracks were first discovered in North America in 1931 in the Dakota Group rocks near Golden, Colorado. This historic specimen (UCM 17614) was named *Ignotornis*. The specimen is early to middle Cretaceous in age.

FIGURE 5.27 Photograph of the first bird tracks discovered in Mesozoic rocks.

were therefore marine and contained no tracks. They also contain virtually no dinosaurs except for the scarce remains of a few carcasses that floated out to sea. Instead, they contain fossils of clams, fish, marine reptiles, and other saltwater animals.

Throughout the strata that represent this long marine interlude—the only one to flood vast areas of the western interior of North America since Carboniferous time, 300 million years ago—one must search for shoreline deposits in order to find tracks. Following Dakota times, the widespread Cenomanian transgression flooded this region so extensively that the shorelines were pushed far to the west and far to the east. Almost certainly dinosaurs and other animals made tracks along these shorelines, but over most of the region such strandline deposits either never accumulated or were not preserved. It was only when the seaway began to shrink a little that the shoreline crept back into the western interior, where it could leave a record of strandline accumulation and animal activity.

In Campanian time, coastal plain deposits were reestablished over large areas of what is now Colorado, New Mexico, Utah, Wyoming, Montana, and Alberta. The large-vertebrate communities at this time were dominated by herbivorous hadrosaurian dinosaurs, popularly known as duck-billed dinosaurs. Hadrosaurs were the descendants of the iguanodontids of the early Cretaceous. These animal communities have become famous, thanks to Jack Horner, who discovered and researched hadrosaur nest sites in Montana. Similar nest sites have been discovered in Alberta, and in both regions the skeletal remains of dinosaurs are abundant in sediments representing inland or upper coastal plain environments.

In the southern part of this region, nesting sites and large accumulations of bones are as yet unknown. This is mainly due to the different nature of the sediments, which at many outcrops represent lower coastal plain deposits characterized by coal-swamp environments. In such environments, where acidity levels are high, bone and other calcareous materials (such as egg shell and mollusc shell) dissolve away, leaving few fossils. However, coal swamp environments allow for the accumulations of fossil plant material and are also conducive to the formation and preservation of tracks. Vegetation accumulates in fetid, reducing environments, where it cannot be broken down by oxidation. Instead, downed vegetation becomes peat and organic-rich mud, eventually—if the conditions are right—becoming coal. In most cases, coal accumulates at or very near sea level in coastal swamps and wetlands. Its occurrence in the fossil record is therefore a good indicator of coastal environments. In such settings, swampy, wetland sub-

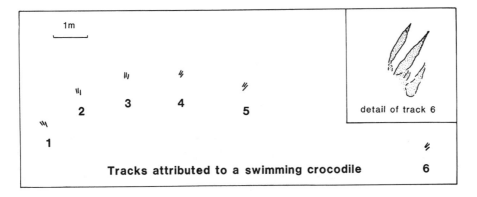

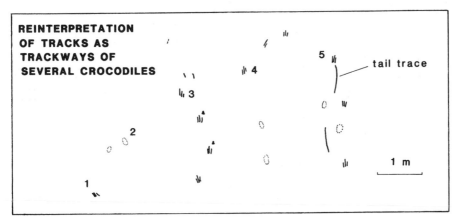

FIGURE 5.28 Initial interpretations of unusual tracks from the Dakota Group as pterosaurian were refuted by later research that showed they were crocodilian in origin. *After Lockley 1991 (top) and Bennett 1992 (bottom).*

strates provide good conditions for track formation. Hadrosaurian dinosaurs evidently found well-vegetated coastal wetlands attractive habitats, because of the abundant food supply, and so frequented these environments.

There are several Cretaceous coal-bearing deposits in the western United States. These can be found in parts of the Dakota Group (Albian-Cenomanian), the Mesa Verde Group, the Laramie Formation, and the Fruitland Formation. All these geologic units are relatively rich in tracks, but none surpasses the Mesa Verde Group in terms of abundance.

Coal Deposits of the Mesa Verde Group

The Mesa Verde Group consists of several formations and extends over parts of at least five states (Colorado, New Mexico, Arizona, Utah, and Wyoming). These strata record the history of coal swamp and coastal plain development on the western shore of the interior seaway as it shifted back and forth during Campanian time. Indeed, geologists use the position of successive coal seams to track sea level changes and to fine-tune their record of sea level fluctuations (figure 5.29).

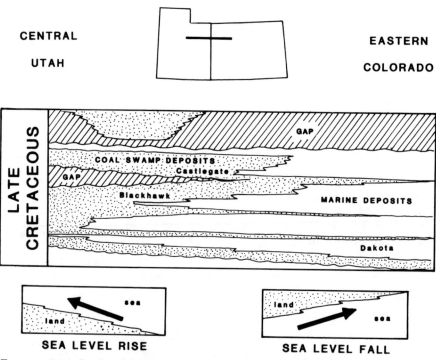

FIGURE 5.29 Sea level fluctuations are recorded in the Cretaceous sequence of coastal plain deposits that mark the western shore of the inland seaway. Stippled sections depict coal swamp deposits; white areas are marine deposits. Many of the coal beds contain abundant tracks. Arrows indicate direction of shoreline movement.

Since the early twentieth century, miners and geologists have been aware of the abundant tracks found at many levels in the coal-bearing Mesa Verde Group. The tracks occur at as many different levels as do the coal seams and attest to prolonged dinosaur activity in these environments. Although tracks are abundant and widespread, they are hard to study. This is because most tracks are exposed only in precipitous cliff faces or in coal mines. In either setting access to the tracks is limited and hazardous. One unfortunate result of this situation is that both lay persons and scientists have speculated freely on the basis of very limited information. Miners have also made a hobby of collecting individual tracks, thereby removing them from the context of trackways and track assemblages to which they belonged. This has reduced their scientific value; interpretations based on isolated specimens are often ill-informed, fanciful, and frivolous. Such circumstances are regrettable because tracks from Mesa Verde coal mines have figured in several publicized debates about trackmaker affinity, speed, and behavior. Usually such debates are hard to resolve without reference to good trackway specimens.

THE CASE OF THE "MYSTERY DINOSAUR"

When first documented by William Peterson in the 1920s, casts of large three-toed tracks observed in a coal mine roof in Carbon County, Utah, were tentatively attributed to a tyrannosaur on the basis of an identification made by W. D. Matthew of the American Museum of Natural History. These very tracks (figure 5.30) were later named *Tyrannosauropus* by the German tracker Hartmut Haubold, simply by reference to Peterson's article. He made no further study of or reference to the original tracks, which have never been relocated. It is very doubtful that the original material was reasonably attributable to a tyrannosaur or indeed to any theropod. The Carbon County coal mines typically yield abundant three-toed tracks of hadrosaurs, so these were probably the trackmakers of the original tracks discovery.

Such ambiguity in track classification is an ominous sign for the science of track identification in the coal mines. Over the next three decades, and indeed until the 1980s, there was little scientific advance in the classification or understanding of tracks from any of the mines in the entire coal belt that extends from southern Wyoming into eastern Utah and much of Colorado. This was unfortunate because several knowledgeable paleontologists were aware of the tracks and presumably had the opportunity to study them—but did not. Among these paleontologists were Richard Swan Lull of Yale University, who had worked extensively with the early Jurassic tracks of New England, and Walter Granger of the American Museum of Natural History, who recognized that the large three-toed tracks were attributable to hadrosaurs. Both these scientists were aware of the work done by Peterson and they were subsequently contacted (around 1930) by Charles Strevell, an enthusiastic amateur who had begun amassing a substantial collection of coal mine tracks. Strevell obtained these as isolated specimens from the Carbon County mines around Price, Utah, and took them to Salt Lake City, where he hoped to compare them with other tracks in order to identify and classify them.

FIGURE 5.30 Purported "tyrannosaur" tracks discovered in a coal mine in Utah in the 1920s. Trackway appears to have been sketched rather than accurately mapped. *After Peterson 1924.*

Unfortunately, Strevell's good intentions were thwarted by his lack of paleontological expertise and the reluctance of professional paleontologists to interpret isolated specimens that had already been removed from the strata where they belonged. Strevell sought the help of both Lull and Granger on how to identify and name the tracks; he was encouraged when they told him the footprints were worth studying. He also consulted Earl Douglass, who had discovered and made initial excavations at what was later to become Dinosaur National Monument. Douglass suggested that Strevell might name the tracks *Dinosauropodes* (meaning, dinosaur footprints). So in 1930 Strevell published a description of several tracks of various shapes and sizes (figure 5.31) and gave them all the ichnogenus name *Dinosauropodes*, with a variety of specific names (for example, *D. osborni* in honor of the famous paleontologist Henry Fairfield Osborn). Strevell's descriptions were inadequate however, and he published them privately rather than in the scientific press. He applied the name *Dinosauropodes* to both three- and four-toed tracks and made no reference to other work on fossil tracks. Thus, most ichnologists regard Strevell's publication as a curiosity rather than a credible scientific contribution.

It was into this scientific vacuum that Barnum Brown of the American Museum of Natural History was able to step with a shrewd but unscientific pub-

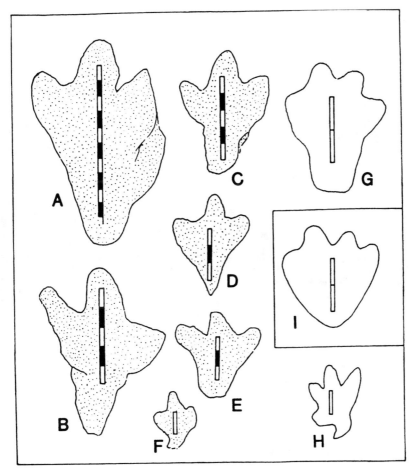

FIGURE 5.31 Various dinosaur tracks from the coal mines of Carbon County in eastern Utah. All tracks (A–H) were initially named *Dinosauropodes* (an invalid name for many reasons). Tracks A–F (stippled) represent tridactyl animals. Tracks G and I are probably attributable to ceratopsians. H may be a double print. Scale bars are shown in 10 cm increments.
Source: Lockley and Jennings 1987 (after Strevell 1932).

licity stunt. Knowing that tracks were reported by Peterson, Strevell, and others, Brown claimed to have found the tracks of a giant "mystery dinosaur" at a mine in Wyoming. This sensational claim, which appeared in *The New York Times* in 1937, was mainly designed as a fund-raising strategy. Moreover, it alerted western coal miners to Brown's interest and encouraged them to report their finds to him. He was soon contacted by Charlie States, who reported giant tracks indicating a five-meter step. States operated the Red Mountain Mine near Cedaredge, Colorado, and was able to offer Brown access to the tracks. Thus, the way was cleared to mount an ambitious excavation to remove the purported giant step. With the help of his invaluable assistant, Roland Bird, Brown initiated a full-scale excavation that kept three shifts of miners working round the clock for three weeks. They excavated a slab seventeen feet in length. It was too large and heavy to extract from the mine with standard equipment, so the crew built a specially constructed mine cart and broke the slab into three manageable sections (figure

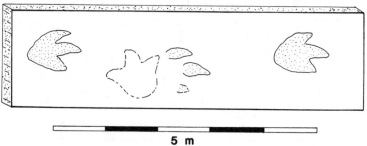

5 m

FIGURE 5.32 Technique for removing a track-bearing sandstone slab from a coal mine. A large track-bearing slab (bottom) excavated from a coal mine in Colorado in the 1930s by Barnum Brown and Roland Bird of the American Museum has long been a source of confusion and controversy. It is simply a hadrosaur trackway in which the middle track (cast) is partly obscured by the tracks of another animal crossing in an oblique direction. *After Lockley 1991.*

5.32) for removal and shipment to the American Museum of Natural History in New York City (where it still resides).

Although the slab showed signs of other tracks between the two widely separated clear footprints, such was Brown's reputation that he was able to publish his claim of a giant five-meter stepper as if it were indisputably proven. Moreover, he played down the obvious hadrosaurian affinity of the tracks by claiming that they were too big to fit known hadrosaurs. This perpetuated the "mystery dinosaur" mystique without suggesting an explanation as to the trackmaker's identity. With the excavation completed, understanding of the significance of these tracks was no better than after Strevell's well-intentioned, amateurish efforts. In fact, one could argue that Brown only added confusion and ambiguity to an already murky situation. And a subsequent generation of enthusiasts were now eager to match Brown's sensational claims.

In the 1950s Al Look, a rock hound and amateur natural history writer from western Colorado, embellished Brown's sensationalism by pointing out that the mystery dinosaur had still not been identified. He casually dubbed it *Xosaurus* without undertaking a formal scientific description. Look further perpetuated the myth of its gigantic characteristics by claiming to have observed a mine where a 4.9-meter (16 foot, 3 inch) step had been measured. As if this were not fantastic enough, he reported another site where the mystery giant had apparently stepped on an animal resembling a crocodile. No evidence was given in support of this intriguing claim.

It was not until the late 1970s that professional paleontologists began to reevaluate the scant data available for dinosaur tracks in coal deposits of the Mesa Verde Group. The renewed interest stemmed largely from a debate about the speeds attained by dinosaurs. There was suddenly a need for trackway data, especially step and stride measurements, to be used for speed estimates that could be derived from a new formula proposed by the English zoologist R. McNeill Alexander. When the "mystery dinosaur" tracks were reevaluated, it was generally agreed that the tracks were those of a large hadrosaurian dinosaur, for which large skeletal counterparts were by then known. However, there was no consensus on step length—even though Brown's slab was available for study. Two authors, Dale Russell and Pierre Beland, claimed that Brown's original observations had been correct—the step was 5 meters—implying an animal running at a speed of about 27 kilometers per hour. Tony Thulborn then disputed this interpretation and noted that the 5-meter "step" revealed an additional partial footprint midway between the clear tracks (figure 5.32). This implied a 2.5-meter step and an animal walking at about 8.5 kilometers per hour. It was not until the 1980s that we obtained new data, which showed that the long, 5-meter steps were apparently overestimates and that Thulborn's interpretation of Brown's American Museum of Natural History specimen was the most reasonable.

Several other studies published in the 1980s demonstrated that tracks from late Cretaceous coal mines are both abundant and diverse. Our own study of a mine near Gunnison, Colorado, revealed numerous hadrosaur trackways with a preferred trend toward the south (figures 5.33 and 5.34). This pattern suggests a gregarious group or herd, as postulated for hadrosaurian dinosaurs elsewhere

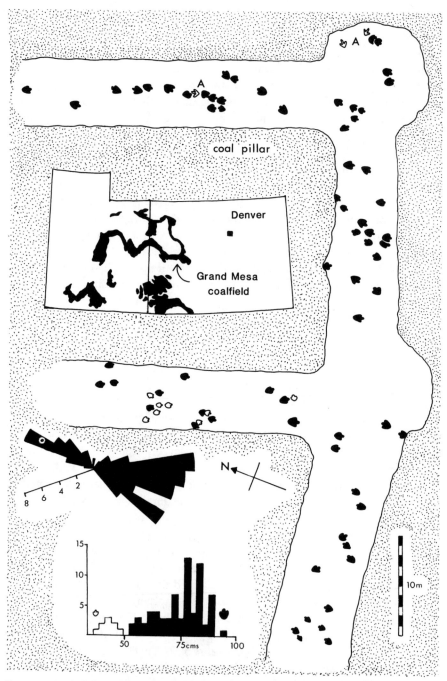

FIGURE 5.33 Map of dinosaur tracks from a coal mine ceiling in the Grand Mesa coal field near Gunnison, Colorado. Inset diagrams show (1) the location of the Grand Mesa coal field, (2) the preferred track orientations, and (3) the size distribution of small (white) and large (black) hadrosaur tracks. *After Lockley et al. 1983.*

on the basis of mass bone accumulations representing thousands of individuals of the same species. It is interesting to note that several early publications on dinosaur tracks from local mines refer to well-beaten trails or pathways, presumably made either by herds or individuals that repeatedly used particular routes. A high density of hadrosaur tracks was also reported by Lee Parker and John Balsley from a mine near Price, Utah, where both the distribution of tracks and tree stumps was mapped (figure 5.35). Some have suggested that the pattern is evidence of a group of browsing hadrosaurs milling around in dense vegetation.

Other reports on Mesa Verde dinosaur tracks have followed the example of Strevell and simply listed the variety of different track shapes observed and collected from various mines. To date, the evidence suggests that large, blunt, three-toed hadrosaur tracks are by far the most common. Large, four-toed tracks presumably attributable to horned dinosaurs (ceratopsians) are also known from isolated tracks, but they are not common and are as yet unknown in trackway sequences from rocks of this age. The same is true for theropod tracks; a variety are known but none, except for the controversial tyrannosaur trackway previously discussed (which we argue is hadrosaurian), have been documented in a consecutive trackway sequence or studied in any detail.

Bird, Frog, Pterosaur(?) and Baby Dinosaur(?) Tracks

In 1989 Lee Parker and John Balsley reported an intriguing concentration of bird-like tracks in a Mesa Verde coal mine near Price, Utah. After consultation with Lockley, they suggested that the footprints could conceivably be attributed to the fossil bird genus *Hesperornis*—which had a very long fourth toe. If this interpretation of a bird affinity for the footprints is accepted, then it implies that *Hesperornis*-like birds, traditionally regarded as divers, inhabited coal-swamp environments as well as marine habitats, where their bones have been found. Moreover, it would imply that the trackmakers could stand up and not merely

FIGURE 5.34 A typical hadrosaur track cast, in late Cretaceous rocks in a coal mine ceiling, Grand Mesa coal field, Colorado. The track is about 90 cm wide. White scratch mark was caused by heavy mining equipment. *After Lockley et al. 1983.*

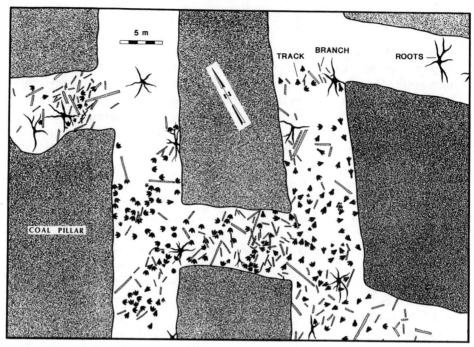

FIGURE 5.35 A map of tracks from a coal mine ceiling near Price, Utah. Notice the abundant tree stump remains. *After Balsley and Parker 1980.*

slide along on their bellies, as is sometimes supposed or observed in modern divers like loons and grebes (whose diving-adapted legs extend rearward of their center of balance). Our own recent studies, however, have suggested another interpretation.

It turns out that the individual tracks are quite similar to tracks described from the Jurassic Summerville and Sundance formations of Utah and Wyoming (figures 4.39–4.41), although the trackway pattern is different. Those Jurassic tracks have now been assigned to pterosaurs, despite being previously regarded as crocodilian. It may seem astonishing that trackers cannot distinguish the tracks of birds, pterosaurs, and crocodilians, but it is worth bearing in mind that, until very recently, few examples of these track types had been found in the Mesozoic, and few occur in clear, simple trackways.

Moreover, the Mesa Verde tracks (figure 5.36) are clearly unusual. They reveal a crude arrangement in pairs (left and right), with the long digit impression directed to the outside of the trackway. The arrangement suggests a trackway width of up to 50 centimeters. This, in turn, indicates an animal with a size and body width that would be consistent with the skeletons of known late Cretaceous pterosaurs.

We freely admit that these track types are difficult to interpret and that their identification as pterosaurian is by no means certain. Despite this uncertainty it is very interesting to note that Parker and Balsley described these tracks as being associated with "many small 1 × 2 cm depressions and several thin (1.5 cm wide) elongate striations," which could be "beak or peck marks made in the mud by the animal as it fed." Beak or dabble marks are known in association with bird footprints (chapter 6), and it is therefore reasonable to infer that the Mesa Verde coal mine marks can be interpreted in the same way—as feeding traces of birds or pterosaurs made in shallow water or along the shoreline.

Having cast doubt on the previous interpretation of these tracks as those of *Hesperornis*-like birds (a dubious interpretation for which the senior author was responsible), it is worth stressing that Steven Robison in 1991 turned up typical tracks of water birds in the Mesa Verde Group (figure 5.37). This new track evidence indicates the presence of birds in these habitats, and it is consistent with many lines of evidence that suggest that birds were a diverse and well-established group by late Cretaceous times. Frog tracks have also been reported from

FIGURE 5.36 Possible pterosaur tracks, with associated beak prod marks, from a late Cretaceous coal mine near Price, Utah. The box encloses a segment of a single trackway. *After Parker and Balsley 1989.*

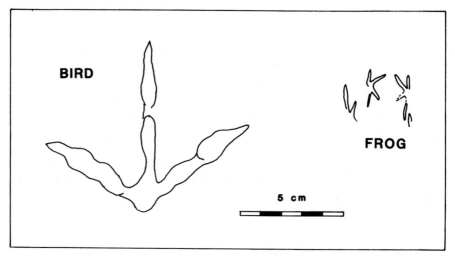

FIGURE 5.37 True bird track from the Mesa Verde Group of Utah, with detail of frog tracks from same locality. *After Robison 1991.*

this same site. Not only are these the first frog tracks ever reported from the Mesozoic (and the world's oldest known frog tracks) but their association with footprints of wading birds raises the intriguing possibility that the water birds preyed on the frogs, as many do today. Perhaps pterosaurs also ate frogs in these environments. Sites such as these indicate the potential for further discovery and an improved understanding of Mesozoic bird and pterosaur tracks.

Currently the best collection of tracks from coal mines is the one on display at the Prehistoric Museum of the College of Eastern Utah in Price. Although the exact strata from which some of the tracks came are unknown, many have come from known levels in the sequence of 22 coal seams of this region. Of some interest are the very small tracks collected by William Wilson (figure 5.38) and thought by some to represent juvenile dinosaurs. If those claims are correct, these are among the smallest dinosaur tracks ever recorded (compare with figure 4.25) and certainly would represent hatchling-sized hadrosaur babies. However, caution should be exercised in the interpretation of these tracks, as we have already shown (in a 1986 publication) that some have been "enhanced" by carving and that almost all these footprints are out of context from the surfaces where they were originally preserved. Moreover, miners who knew Wilson have speculated that he deliberately looked for anything he thought might be a small track, and as a result accumulated a biased sample. These observations suggest a need for further study of in situ surfaces, where a complete picture of the track record can be obtained before interpretations are proposed. Further study of accessible mines promises to improve the understanding of track diversities, paleoecology, and biostratigraphy in this area.

UNDERGROUND HAZARDS

Mining is a hazardous profession. Every mining community has known tragic incidents of death, injury, or disease resulting from explosions, cave-ins, and exposure to dust and bad air. In the mining communities of the western United States, anecdotes and apocryphal stories circulate about dinosaur tracks that fall out of mine ceilings. Given that such large sandstone casts can weigh hundreds of pounds, they clearly pose a threat to life and limb.

Anecdotal as these stories may be, they do contain some truth. Most tracks in the roofs of coal mines represent footprint impressions in the peaty coal that were later filled in by sand. In come cases the sand only partly filled the tracks before finer-grained mud and silt accumulated to produce the overlying layer (figures 1.10 and 5.39). This then created a sandstone pocket or protrusion that could easily detach from the overlying layer. Miners routinely bolted these protrusions to the ceiling to prevent them from dislodging.

Although we know of no documented examples of tracks that fell and killed miners, we do know of a fatality caused by a dinosaur track. In March of 1969 a miner working near Hayden, Colorado, ran to stop a coal car that began to roll away from him. In doing so he struck his head on one of the many tracks that protrude 3–4 inches below roof level. Although he remained conscious and was able to make his way home, he sustained a fatal spinal injury and died ten days later in a Denver hospital.

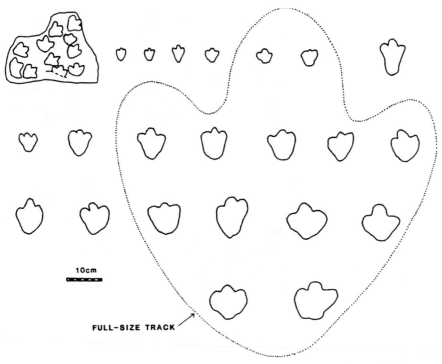

10cm

FULL-SIZE TRACK

FIGURE 5.38 Diminutive tracks from coal mines in the area of Price, Utah, may indicate the activity of juvenile hadrosaurs. Dotted outline indicates the size of an adult hadrosaur track for comparison. *After Lockley 1986.*

In the book *Bones for Barnum Brown*, author Roland Bird described the hazards of subterranean fossil hunting. In his judgment, excavations of large trackways that he undertook in mines have been relatively safe compared to what he regards as his own foolhardy excavations of plant fossils in decidedly unsafe locations. In earlier times, as miners excavated along coal seams they removed roof supports, thus allowing the roof to collapse under overburden pressures. It was in these marginal areas that Bird reported collecting some prize plant fossils.

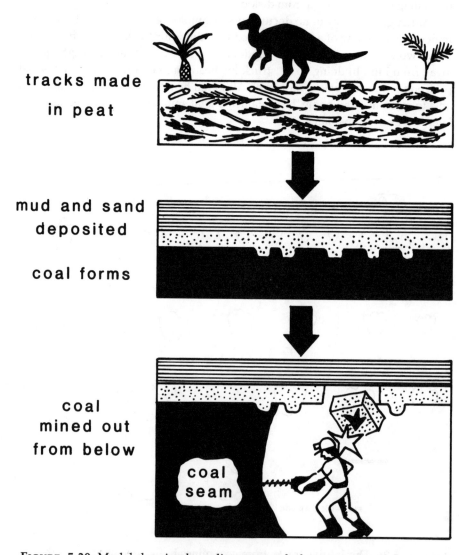

tracks made
in peat

mud and sand
deposited

coal forms

coal
mined out
from below

coal
seam

FIGURE 5.39 Model showing how dinosaur tracks became preserved as isolated sandstone casts that easily detach from coal mine ceilings after the coal has been excavated. Such detachable tracks pose a hazard to miners.

Although adventurous, Bird was normally a cautious man, but in this case his foolhardy quest for fossils could easily have cost him his life.

Rare Tracks of the Laramie Formation

Along the Colorado Front Range, particularly near Denver, outcrops of late Cretaceous sandstone and shale have been extensively mined both for refractory clays, used in brick making and ceramics, and for localized coal deposits. These deposits have also become well known for the abundant plant fossils they contain. However, not until the mid 1980s did it become clear that these sandstones and shales—known as the Laramie Formation—were rich in fossil footprints. What is more, the formation is rich in a kind of dinosaur track that is not known to be abundant anywhere else in the world. And it has offered up the only tracks yet discovered of an extinct aquatic reptile.

The formation also contains some dinosaur bone attributable to horned dinosaurs, or ceratopsians, of which the most famous is *Triceratops*. Indeed, the Laramie Formation is part of what is known as the *Triceratops* biozone, which also extends into the overlying Denver Formation. The latter contains the famous Cretaceous-Tertiary boundary that marks the extinction of the dinosaurs and several other groups of Mesozoic animals.

Given the abundance of ceratopsian remains in the Laramie Formation and other latest Cretaceous deposits of the region, it is not surprising that many levels in this formation contain tracks of ceratopsian dinosaurs (figure 5.40). However, before the discovery of fossil footprints in this formation, ceratopsian footprints were virtually unknown. The Laramie tracks are thus significant as the first substantial record of the footprints of horned dinosaurs and the first formation to yield clearly recognizable trackways (figure 5.41).

Paleontologists have debated for several generations as to whether ceratopsians stood erect like a rhinoceros or had their elbows sticking out sideways from the body in an ungainly sprawling or semierect posture (figure 5.42). The bones suggest, to some degree, that ceratopsians could not easily articulate their front limbs in an erect posture. What do the trackways indicate? As shown in figure 5.41, the front footprints are situated in front of the hind foot tracks. This suggests a relatively erect posture, as seems to be typical of all major groups of dinosaur, including the quadrupedal varieties. In short, ceratopsians appear to have walked erect, like most other dinosaurs.

The rarity of ceratopsian tracks, prior to their discovery in the Laramie Formation, flamed a debate about the scarcity of certain track types. The debate turned on the puzzle that the tracks of most large quadrupedal ornithischian dinosaurs—especially ceratopsians, ankylosaurs, and stegosaurs—are rare or unknown, whereas their skeletal remains are common, even abundant in places. This puzzle is all the more intriguing when one considers that the rarity of these tracks cannot be attributed to a trackmaker being so small that it fails to depress the sediments; these particular trackmakers were heavy animals. At present there are still no well-accepted examples of stegosaur tracks and only a few convincing examples of ankylosaur tracks. Few ceratopsian tracks are known from

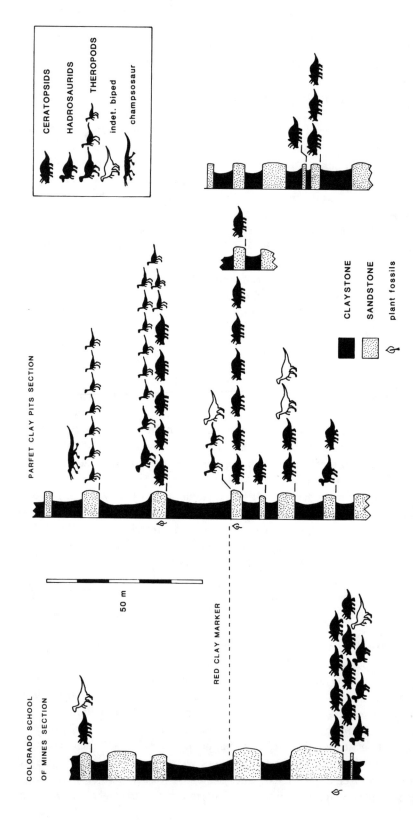

FIGURE 5.40 Tracks of ceratopsians and other dinosaurs are found at several different levels in the late Cretaceous Laramie Formation near Denver, Colorado. Each animal silhouette represents one individual trackway. *After Lockley and Hunt 1994 and in press.*

deposits other than the Laramie Formation. How can this track sparsity be explained?

One appealing argument for the scarcity of large, quadrupedal tracks of the ornithischian group of dinosaurs is that the trackmakers generally, or frequently, avoided wet habitats or environments where tracks are most often preserved. They were upland or dryland animals rather than lowland or wetland forms. Several lines of evidence suggest that stegosaurs were upland or dryland species, but the evidence for ankylosaurs and ceratopsians being so is less persuasive. In fact the evidence from the Laramie Formation, where tracks are quite deep, suggests a locally wet area associated with many river channels. Perhaps, in part, certain kinds of tracks are puzzlingly scarce simply because paleontologists have not looked enough for rare tracks, or they have failed to look in the right places. As more tracks are found, it becomes easier to establish which habitats particular

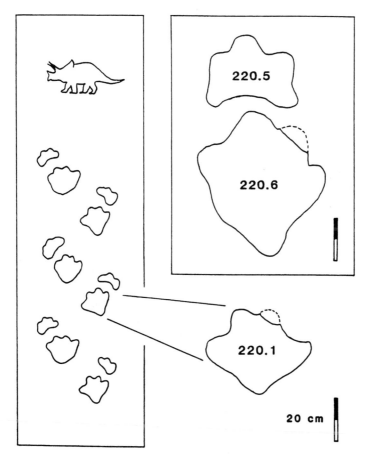

FIGURE 5.41 Ceratopsian tracks and trackways from the Laramie Formation. Compare with Figure 1.5 to see match between track and foot skeleton. *After Lockley 1986 and 1988; Lockley and Hunt 1994 and in press.*

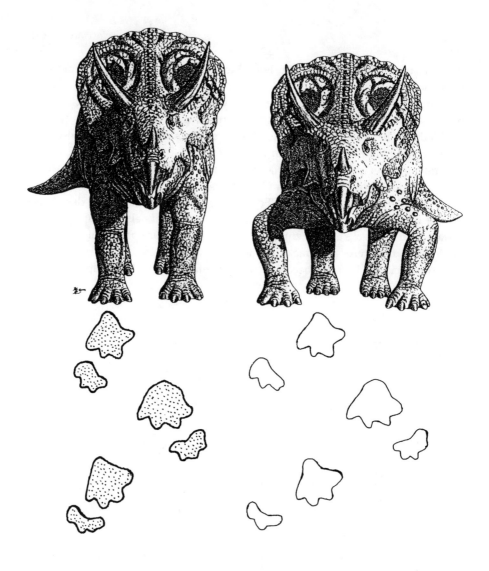

RIGHT   **WRONG**

FIGURE 5.42 Reconstructions of ceratopsian posture showing alternative theories of erect versus semierect front limbs. It is now generally believed that ceratopsians walked with erect front limbs. *Artwork courtesy of Pat Redman.*

trackmakers frequented, and whether a perceived rarity of tracks is real or simply an artifact of human effort.

The Laramie trackways show that ceratopsians shared their habitats with a few large hadrosaurs and a number of other three-toed trackmakers interpreted as theropods (figure 5.43). The three-toed trackmakers exhibited a considerable size range, from turkey-sized species to a very large *Tyrannosaurus*-sized trackmaker (see box and figures 5.44–5.45). The tracks therefore indicate a high diversity of theropods at this time presumably all inhabiting different niches.

We discovered one other distinctive track type in the Laramie Formation near Denver (figure 5.46). It is of a small, five-toed creature that probably represents a champsosaur. Champsosaurs were primitive, aquatic, lizard-like reptiles that resembled crocodiles and that were relatively common in late Cretaceous times, though their tracks have never previously been reported. This is perhaps because

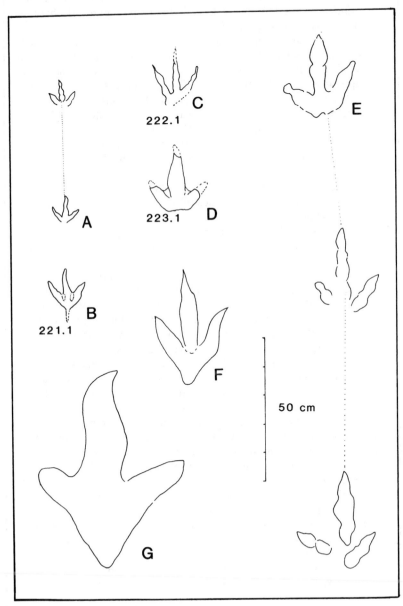

FIGURE 5.43 Tridactyl tracks in the Laramie Formation indicate a variety of theropods. The largest theropod track may be attributable to *Tyrannosaurus*. *After Lockley 1988; Lockley and Hunt 1994.*

Trex Trax?

Where are the tracks of *T. rex*? Following the recent discoveries of *Tyrannosaurus rex* skeletons in Montana and North Dakota—not to mention the release of the movie *Jurassic Park*—*T. rex* is enjoying even greater popularity than before. The Montana skeleton was featured in Jack Horner and Don Lessem's new book, *The Complete T. Rex*, and the North Dakota specimen named Sue has gained notoriety after being impounded by the FBI, owing to an ownership dispute. All this has prompted many to ask about the track evidence.

Although *Tyrannosaurus* is a well-known dinosaur, its remains are quite rare. The same appears to be true for *Tyrannosaurus*-like tracks. Although one of the aforementioned trackways from the Mesa Verde Group was attributed to a large tyrannosaur, there is really little evidence to support this assignment. In fact, the tracks are too old to have been made by *T. rex*, and the trackway is probably attributable to a hadrosaur. Overall, there are very few *Tyrannosaurus*-like tracks known from strata of the same age as strata in which *Tyrannosaurus* bones occur. Until recently, the lone Laramie Formation track pictured in figure 5.43 was one of the rare examples of a possible *Tyrannosaurus* track in latest Cretaceous (Maastrictian) deposits of the right age. Because the Laramie track is now buried, at a filled-in quarry site, it is no longer accessible for study.

At the time of writing, however, we have just identified and described a much better preserved example of a probable tyrannosaur track, discovered in the Raton Formation of New Mexico by Chuck Pillmore (figures 5.44 and 5.45). This track is clearly different from typical hadrosaur tracks and has many features indicative of theropod tracks. These features include a length (85 cm) that exceeds the width, a distinctive hallux impression, and an indentation behind digit II. *Tyrannosaurus* is the only genus of dinosaur known to have been present in North America in the latest Cretaceous that could have made a track of this shape that was this big. Unfortunately, this is the only track of its kind we have been able to find on this particular Raton Formation rock outcrop, and although the track is clear and deep, there is no sign of the next track for a distance of almost 3 meters, to the edge of the outcrop. The step length must, therefore, have exceeded this distance. The track is found only a few meters below the K/T boundary, and so could be the youngest known evidence for *Tyrannosaurus*—as well as the only track of its kind currently known in the world. We named the track *Tyrannosauripus pillmorei* in honor of the discoverer.

At the risk of repeating ourselves, the lack of track evidence is probably in part due to a lack of systematic searching. The recent discoveries of *T. rex* skeletons have shown how much is yet to be learned about this "king of the dinosaurs." We are hopeful that the future will indeed bring the discovery of a well-preserved *T. rex* trackway that will provide solid evidence for making estimates of speed, and possibly new information about foot shape, erectness, gait, and so on.

of their aquatic habits. Tracks of aquatic animals are generally rare, compared with their terrestrial contemporaries. Caution is always advisable, as in this case, when claiming that a newly discovered track type should be assigned to a particular group of trackmakers. After all, trackers have confused the footprints of pterosaurs and ornithischians with those of turtles and crocodiles, and the debate about what prosauropod tracks look like has lasted for more than a century.

In the case of our alleged champsosaur, we use two arguments to support our claims. First, the size and shape of the tracks fit extremely well with known champsosaur foot skeletons. Second, champsosaurs are common fossils in western North America at this time. Champsosaurs were an interesting and unusual

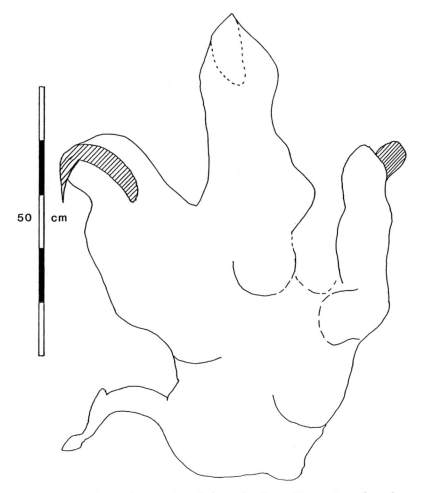

50 cm

FIGURE 5.44 A huge theropod track from the Raton Formation of northern New Mexico, discovered in 1983 but identified in 1993. It is probably attributable to the species *Tyrannosaurus rex*, because no other theropod of that time had feet big enough to make this track. This is the first reliable report of any *Tyrannosaurus* tracks from rocks of the right age.

Figure 5.45 Fiberglass replica of the *T. rex* track from the late Cretaceous of northern New Mexico. Track is 85 cm long.

group of animals. They were very crocodile-like in appearance and presumably in habit, but they were lepidosaurs (related to lizards) rather than archosaurs (the group to which true crocodiles belong). Unlike the non-avian dinosaurs and certain other groups of large reptiles, champsosaurs successfully survived the late Cretaceous mass extinction—only to die out early in the Tertiary period.

Given the abundance of late Cretaceous tracks in the Laramie Formation (and in western Canada), the paucity of track reports in bone-rich strata of the same age in Wyoming and Montana is puzzling. Formations such as the Two Medicine, Judith River, and Hell Creek in those states are famous for an abundance of skeletons of late Cretaceous dinosaurs. Even dinosaur eggs and nest sites have been discovered there in abundance. One would expect dinosaur tracks to be quite common in these formations as well. So why not?

The paucity of track reports in Wyoming and Montana is probably due to lack of systematic exploration and, perhaps in some cases, unfavorable conditions for track preservation. For example, we know that dinosaurs must have walked in and out of the nesting colonies recently made famous by the work of Jack Horner, but where are their tracks? Perhaps the ground was compacted or trampled to the point where individual tracks could not be impressed, or at least not preserved in recognizable form.

In the first edition of this book we predicted that tracks would be found abundantly in the upper Cretaceous of Wyoming and Montana, and noted that possible bird, crocodile, and tortoise tracks had been reported but not documented from the Two Medicine, Judith River, and Harebell formations. The Harebell tracks have since been named *Saurexallopus*. We now know that large theropod tracks have been found in the Hell Creek Formation of western South Dakota, and that a diverse assemblage of dinosaur and bird tracks in known from the bone-rich Lance Formation of Wyoming. Tracks from this site include hadrosaur footprints with skin impressions, theropod tracks including *Saurexallopus*, and four different track types attributable to waterbirds. Similar new finds have been made in western Canada.

Raton Formation Tracks and the Cretaceous Extinctions

One of the biggest debates in paleontology has revolved around the extinction of the dinosaurs. Numerous hypotheses have been proposed, postulating single or multiple causes. The debate has attracted unprecedented public attention, not least because of the dramatic nature of events inferred to have brought the Mesozoic era to a close. This ending of the Cretaceous period and beginning of the Tertiary period or "Age of Mammals and Birds" is referred to as the K/T boundary—K for Cretaceous (C being used for Carboniferous) and T for Tertiary.

In many areas the actual stratigraphic boundary is marked by a layer of clay rich in the element iridium. Iridium is extremely rare on the surface of our planet but it is found in somewhat higher concentrations in some meteorites and deep within Earth. This iridium layer, also sometimes referred to as the "magic layer"

FIGURE 5.46 A track of a champsosaur, a kind of aquatic lizard, from the Laramie Formation near Golden, Colorado (center). This tracks fits well with the known anatomy of champsosaur feet (left and right). *After Eriksen 1972; Lockley and Hunt 1994 and in press.*

or "boundary clay," is often associated with microscopic grains of shocked quartz and with soot. This layer thus undoubtedly represents a major catastrophic event, and a majority of scientists have now accepted that a meteorite, asteroid, or other extraterrestrial "bolide" collided with Earth. A few, however, propose that the iridium and shocked minerals may have been the result of volcanic events here on Earth, and there is indeed good evidence for massive outpourings of lava, known as the Deccan Traps, in the subcontinent of India at the time of the K/T boundary. Some impact proponents reply that the Traps were themselves caused by an impact.

For over a decade the extinction debate between the impact advocates and those who favor terrestrial causes has swung back and forth. The impact advocates recently had their position strengthened by the identification of a large buried crater of the right age on the Yucatan Peninsula of Mexico. Regardless of the merits of competing hypotheses, or of the new evidence that will undoubtedly come to light in the future, the best approach to understanding K/T events and their affect on the biota is careful analysis of the geological and paleontological evidence that is found in rocks below, in, and above the boundary layers. Much effort has already been expended on analyzing the chemistry and characteristics of these rocks (including fossil soils), the changes in microscopic plankton and invertebrate remains in marine deposits, and the exact positions—above or below the boundary—of dinosaur bones and those of other vertebrates, plants, and pollen in continental deposits. However, little attention has been paid to the utility of tracks in understanding the K/T event.

Tracks are useful in dating various early Mesozoic rock units, and the track record clearly shows the dramatic faunal turnover that occurred at the Triassic-Jurassic boundary. As recently discovered evidence reveals, tracks at the K/T boundary shed significant light on the extinction of the dinosaurs. To date, the youngest known Cretaceous tracks anywhere in the world are a small assemblage of hadrosaur tracks, which we found atop a sandstone only 37 centimeters below the iridium layer at a site in the Raton Formation near Ludlow, Colorado (figure 5.47). The next youngest tracks, also hadrosaur, are located 59 centimeters below the boundary at the same site.

The rocks in the track-bearing layers of the Raton represent coal swamp or peat bog environments and are rich in plant remains, including leaf impressions and fossil pollen and spores. The tracks indicate that hadrosaurs were alive and well not long before the event that caused the iridium fallout. Based on sediment accumulation rates in similar environments today, 37 centimeters of mud and silt accumulation in a peat bog probably represents only a few thousand years—a geologically short period of time.

We have recently investigated two new track localities straddling the K/T boundary, 25 kilometers south of Ludlow, near Starkville, on the outskirts of Trinidad, Colorado. These two localities, like the one at Ludlow, are in the Raton Formation (figure 5.47). These new sites reveal the trackways of at least ten vertebrates. Hadrosaur and ceratopsian tracks are found within one to two meters below the boundary. And as we already discussed, our discovery of the first near-certain *Tyrannosaurus* track was also in the Raton Formation; the track was

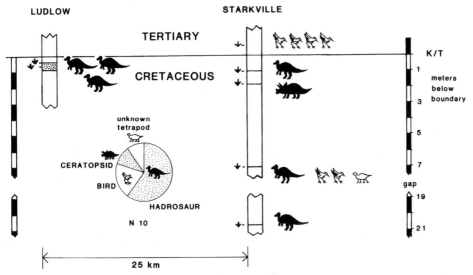

LUDLOW   STARKVILLE

TERTIARY

CRETACEOUS

K/T

1

meters
below
boundary

3

unknown
tetrapod

CERATOPSID

BIRD

HADROSAUR

N 10

5

7

gap

19

21

|← 25 km →|

FIGURE 5.47 Tracks near the K/T boundary in southeastern Colorado.

about 20 meters below the boundary. Based on our survey, only the tracks of birds are found just above the boundary.

The track data from Colorado combine to provide us with a census of ten of the last individual animals known to have lived in the Cretaceous (figure 5.47), and this from the Raton Formation from which no vertebrate skeletal remains have yet been reported. When taking all the track data into consideration, our preliminary synthesis suggests that hadrosaurs were the most common animals in terminal Cretaceous time, in this region, but that ceratopsians, tyrannosaurs, birds, and other unknown trackmakers were also present. The Raton Formation surely has the potential to offer up far more and, notably, diverse footprint faunas both above and below the K/T boundary. This area may indeed be an ichnological Rosetta Stone for unlocking some of the paleontological mysteries surrounding the K/T boundary event.

The discovery of the hadrosaur, ceratopsian, and tyrannosaur tracks close to the K/T boundary sheds important light on a long-standing debate about the demise of the dinosaurs. Paleontologists and other earth scientists have debated whether dinosaurs died out gradually or abruptly, basing their conclusions largely on the distribution of skeletal remains near the K/T boundary. Despite extensive searching, no one has yet found much in the way of complete skeletal remains in sediments close to the K/T boundary—at least not in the few meters immediately below. Taken at face value this would suggest that dinosaurs died out, or at least declined significantly, before the K/T event. The presence of *isolated* skeletal remains in sediments immediately below the K/T boundary suggests to some that the dinosaurs survived until the K/T event, and then died abruptly. Dinosaur remains (mainly teeth) have even been found in sediments above the K/T boundary layer, and the possibility of Tertiary dinosaurs, or K/T event survivors, has recently been widely discussed.

The problem with isolated skeletal remains, both above and below the boundary, is that they can be explained by reworking. That is, they may be nothing more than remnants of dinosaurs that were washed out of older sediments and then reburied in younger deposits. This possibility brings us back to our starting point, where we noted that there are no complete (nonreworked) dinosaur skeletons yet known from immediately below (or above) the K/T boundary layers. This fact leads to the idea that there may be a gap in the fossil record of terrestrial vertebrates, sometimes referred to as the "three-meter gap." In the recent book, *The Complete T. Rex*, authors Jack Horner and Don Lessem assert, "We don't find any dinosaurs within 100,000 years of the iridium band. In eastern Montana there is no dinosaur skeleton closer than three meters below the K/T boundary line." Although this represents a lack of evidence (or negative evidence), it has been used by some paleontologists to argue for a gradual decline of dinosaurs. By contrast, rapid extinction advocates suggest other explanations for the putative gap—such as adverse soil chemistry that caused dinosaur skeletal remains to dissolve. All these arguments are to a large extent redundant in the face of the track evidence, which proves that at least some dinosaurs were alive and well toward the end of the three-meter-gap phase. If three meters represents approximately 100,000 years, according to some estimates, then the 37-centimeter gap is an order of magnitude less (i.e., 10,000 years). Some geologists would estimate an order of magnitude less than this, or only 1,000 years. Clearly, footprints have helped close the gap significantly. More and more, footprints are proving their worth in paleontological investigations.

Ten years ago dinosaur tracks were still regarded as relatively rare and insignificant components of the fossil record. No one would have predicted that they would contribute specific and precise chronological information to important debates about extinction. Few paleontologists would have seriously considered their utility or even their potential in studies of dinosaur populations and communities. And who would have predicted the existence of trampled zones that extend for tens of thousands of square miles, thus giving an estimated data base of millions of footprints?

Almost a decade ago our research group compiled a 56-page descriptive booklet on dinosaur tracksites of the West, covering most of what we knew at that time and venturing very little in terms of the significance of the tracks for interpreting dinosaur biology. By contrast, it has been a challenge to know what to include and what to leave out of these last three chapters—and where to draw the line in pursuing fruitful lines of interpretation. In our final chapter we will discuss some of the broader implications of this burgeoning data base, but first, and for completeness, we need to look at the Cenozoic age of birds and mammals. Here we will see an echo of the poor status of knowledge of and interest in dinosaur tracks only ten years ago. We thus approach our survey and analysis of the Cenozoic with the confidence that in another decade or so that track record will no longer be regarded as either insignificant or sparse.

Sloth, Mammoth, and Herons make tracks on the shores of a Pleistocene lake. Based on track from the Carson City Jail site, Nevada, once written about by Mark Twain.

6

THE AGE OF BIRDS AND MAMMALS: THE CENOZOIC ERA

The initial aim of my study was to determine to what extent, if any, the living genera and species of salamanders in a particular region may be distinguished by their trackways alone. Results could then be applied effectively to . . . fossil trackways . . . in Tertiary strata.

FRANK PEABODY (1959)

■   The Cenozoic era (66 million years ago to the present) is commonly known as the Age of Mammals, though some have also referred to it as the Age of Birds, or even the Age of Social Insects, because all these groups of animals were spectacularly successful during this time. The Cenozoic is divided into the Tertiary period (66–1.6 million years ago), which is subdivided into five epochs (figure 6.1), and the Quaternary period, which more or less corresponds to the Great Ice Age (Pleistocene epoch, 1.6 million to 10,000 years ago), but also includes the postglacial Holocene or Recent epoch.

Traditionally, a large majority of vertebrate paleontologists in North America have focused attention on studies of Tertiary mammals from western states, such as Wyoming and Nebraska. Much less attention has been devoted to the skeletal remains of other vertebrate classes (birds, reptiles, and amphibians) or to tracks. Because tracks have generally been ignored or overlooked, ichnologists studying the Cenozoic are at a disadvantage. However, as for most of the vertebrate track record, this lack of knowledge is not because of a shortage of tracks. Ongoing work continues to reveal many new, or previously undocumented, sites, and we anticipate that the summary presented here (figure 6.1) will need updating in the near future.

Although the Tertiary period makes up most of the Cenozoic era, and although it is much longer than the Quaternary period, it is no longer than typical Mesozoic and Paleozoic periods. It is sometimes divided into the older Paleogene and younger Neogene "subperiods" or "superepochs." These are

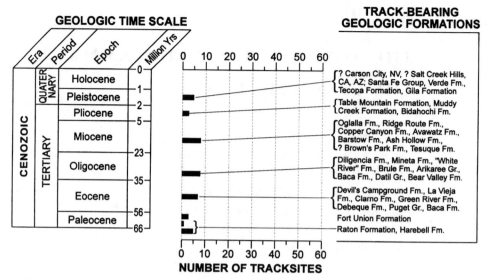

FIGURE 6.1 Summary of known track-bearing formations of Tertiary and Quaternary age.

sometimes regarded as the ages of archaic and modern mammals, respectively. Among the hoofed mammals, or ungulates, the odd-toed varieties dominated in the Paleogene, while the even-toed varieties rose to prominence in the Neogene. The odd-toed forms (1, 3, or 5 toes) are known as perissodactyls and include horses, tapirs, and rhinos. The even-toed forms (2 or 4 toes), known as artiodactyls, include the cloven-hoofed camels, sheep, pigs, deer, elk, bison, and hippos—many of which are cud chewers, or ruminants. The order Carnivora (dogs, cats, bears, etc.) accounts for most of the other common mammalian trackmakers known from the Cenozoic. Other important mammal groups, like the rodents and primates, have, unfortunately, left tracks at few sites.

In chapters 1 through 5 we introduced a number of Latin names (ichnogenera and ichnospecies) that mainly applied to tracks of various reptiles and amphibians. The same taxonomic rules apply for birds and mammals in the Cenozoic; Latin names are used liberally, but there are some differences. Attempts have been made to incorporate familiar modern names into the formal track terminology used for distinctive groups of tracks. This scheme, proposed by Oleg Vialov, a Russian tracker, refers to carnivore tracks as "Carnivoripeda," with "Bestiopeda" and "Felipeda" for dog and cat footprints. Cloven-hoofed artiodactyl tracks are referred to in various categories, such as "Gazellipeda" (gazelle tracks) and "Pecoripeda" (peccary or pig tracks). "Proboscipeda" has been used for elephant tracks and "Avipeda" for bird or avian tracks.

Although ancient tracks have formal names distinct from the trackmaker, modern tracks do not. They are simply referred to as dog tracks, camel tracks, and so on. This scheme is thus something of a compromise between plain English and the lack of formal names for modern tracks, on the one hand, and the universal practice of classifying and naming organisms and fossil footprints, on the other. Formal names have been used only sparingly in the few studies of western Cenozoic tracks that have been published to date.

Paleocene Tracks and the Survivors of the Mass Extinction

The first epoch of the Tertiary period, the Paleocene, lasted from 66–59 million years ago. By an interesting coincidence, the oldest known Paleocene tracks occur in the Raton Formation, in sediments that are only about 60 cm above the K/T iridium layer. These tracks are of several birds (figure 6.2); they were only recently found at a site near Trinidad, Colorado, not far from the youngest Cretaceous tracksite at Ludlow. Kirk Johnson of the Denver Museum of Natural History was the discoverer. In conjunction with Johnson, we have since found tracks of mammals and other small trackmakers of unknown affinity at the Trinidad site.

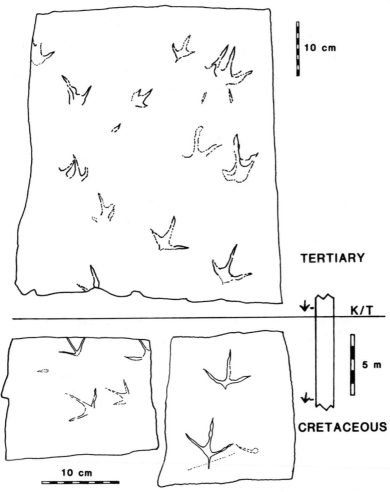

10 cm

TERTIARY

K/T

5 m

CRETACEOUS

10 cm

FIGURE 6.2 Bird tracks from just below and just above the K/T boundary, near Trinidad, Colorado. *Courtesy of Kirk Johnson, Denver Museum of Natural History.*

As no skeletal remains have been described from the Raton Formation, this track evidence adds substantially to the record of vertebrate animals known at K/T boundary times. The tracks below the boundary confirm that, at least in this area, dinosaurs and birds were active before the iridium fallout. The tracks just above the boundary indicate that other birds, plus mammals, were active afterward.

In general, however, fossil footprints are poorly known and little studied in Tertiary deposits of any age—and this is particularly true of the Paleocene epoch. From the fossil record, it seems that no vertebrate with a body weight of more than about 25 kg (about the size of a medium-sized dog) survived into the Tertiary. Thus, the early Paleocene was characterized by a fauna in which large animals were not represented. This fact may help to explain the scarcity of Paleocene tracksites in the western United States, as tracks of light-weight animals are generally rare.

The only Paleocene tracksites that have been the subject of published studies are two sites in the Fort Union Formation—the first in Montana and the second in Wyoming. At these sites the predominant trackways are those of amphibians and birds (figure 6.3). The Montana site, where only amphibian tracks have been found, was discovered in 1908. Charles Gilmore published the first description of the site in 1928. He named the tracks *Ammobatrachus montanensis*, using an ichnogenus name he had recently coined for Permian tracks (200 million years older) from the Grand Canyon.

In 1954 Frank Peabody, who is well known for his studies of Triassic tracks and his experiments with modern salamanders, reexamined these Paleocene amphibian tracks. He renamed them *Ambystomichus montanensis* to stress their affinity to the modern family of newts known as the ambystomids. Gilmore had pointed out in 1928 that his was the first report of any Paleocene tracks from anywhere, but Peabody went one step further. Peabody showed that these tracks represented the earliest track record of the order of tail-bearing amphibians known as Urodela. (Note the tail drag mark in figure 6.3). He also did ichnologists a favor by getting rid of a Permian track name in the Paleocene literature. Although there is nothing wrong in principle with having the same name for tracks of widely different ages if they are identical tracks, this was not the case with the Montana specimens. Peabody's revision therefore avoids the confusion and incongruity of an implied close relationship between the Permian and Paleocene trackmakers.

In 1986 Kirk Johnson discovered more amphibian tracks in the Fort Union Formation at the Elk Basin anticline, near the Wyoming-Montana state line, south of Billings. Johnson also recorded two types of bird tracks, including a large variety (Type 1) that shows partial webbing and a hallux, and a smaller variety (Type 2) without web impressions (figure 6.3). Both bird track types were made by medium-sized wading birds. Johnson reconstructed the environment in which the trackmakers were active and concluded that the tracks were made on a sand bank near a main channel of a stream. He also observed various insect trails and leaves and suggested that while the insects fed on decomposing vegetation, the birds and amphibians may have fed on the insects.

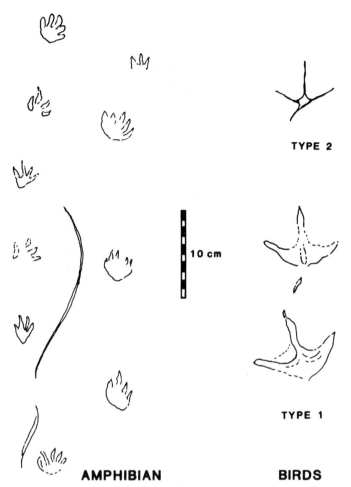

TYPE 2

10 cm

TYPE 1

AMPHIBIAN   **BIRDS**

FIGURE 6.3 Paleocene amphibian tracks from Montana (after Gilmore 1928) and two types of Paleocene bird tracks from Wyoming (after Johnson 1986).

The absence of mammal tracks at both of these sites is noteworthy. The absence may reflect the impoverishment of mammal faunas early in the Tertiary, but it probably also reflects lack of study. Some poorly preserved tracks that are possibly mammalian were reported from the Paleocene of Alberta in 1928. But besides this example, we know of only a few other reports of Paleocene mammal tracks from western North America.

Another discovery of Paleocene tracks, from a site near Baggs, Wyoming, has yielded an assemblage of bird tracks along with the dog-like tracks of an early mammalian carnivore known as a creodont, which we describe here for the first time. Segments of this putatively 60-million-year-old trackway on display at the Raymond Alf Museum in Claremont, California, show two creodonts going in exactly opposite directions (figure 6.4), perhaps parallel to a shoreline. Close inspection of the trackways shows that they reveal overprinting and that the hind foot was placed almost directly on top of the front footprint. This pattern is typical of mammalian walking gaits and sometimes results in tracks that reveal more that five toe impressions.

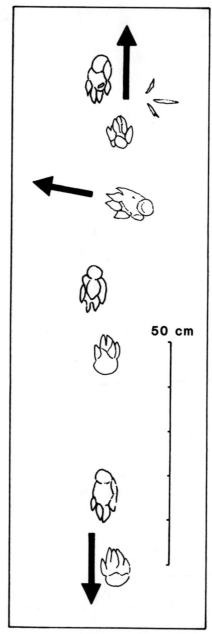

FIGURE 6.4 Dog-like creodont tracks that represent an early Tertiary carnivore. Note two trackways with opposite orientations.
Traced by Don Lofgren from a specimen in the Raymond Alf Museum.

If one is dealing with isolated tracks, it is possible to interpret such tracks as rare examples of polydactyly—a condition sometimes seen in modern cats, where an animal really does have six or even seven toes. Another interpretation is that the tracks are "syndactylous," meaning that two or more digits are partly fused together in the same skin or integument. This condition is seen in some living marsupials, a group that was known to exist in early Tertiary time but that is now mostly restricted to Australia and South America. Fossil footprints attributed to syndactylous trackmakers have been reported from Eocene beds in Texas. Intriguing as these possible explanations are, we believe they are improbable and that the trackways simply show evidence of overprinting. Moreover, overprinting is by far the simplest explanation. Probability dictates that overprinting will be seen in a large proportion of mammal trackways but that polydactyly and syndactyly will be rare.

Is the Decimal System Based on Five Fingers and Five Toes?

Polydactyly, especially in the realm of tracks, is rare. But some of the first tetrapods to ever walk on land were themselves polydactylous. Skeletal fossils of Devonian amphibians from Greenland and Russia reveal 6, 7, or even 8 toes. These foot shapes would surely result in very distinctive tracks if any were known for these creatures. However, Devonian amphibians were evidently the exception to the rule; after this early experimentation with polydactyly, all subsequent tetrapods appear to have had only five or fewer digits.

If these early creatures had not settled on the five-fingers-plus-five-toes blueprint, the whole tetrapod lineage might have developed with a 6-, 7-, or 8-toed design. If these higher digits had carried through to primates, and our own species, we might be sitting today in front of keyboards designed for 12, 14, or 16 fingers. Indeed, our whole concept of numbers might be different, based on larger even numbers rather than on the decimal system. There might have been 120, 140, or 160 pennies in a dollar or a pound, and correspondingly more centimeters in a meter, or grams in a kilogram. Olympic athletes might have run the 160 meter dash, and all other races on longer tracks.

Since the time of multi-digited amphibians the trend in tetrapods has been either to retain five digits or to reduce the number. Thus, had primates (and humans) ended up with less than five digits on each limb, our numerical system might have been based on 8, 6, 4, or even 2. There are many examples of tetrapod lineages that lost some of their digits—becoming 4-toed, 3-toed, 2-toed, or 1-toed, as in the case of hippos, birds, sheep, and horses, respectively. However, no lineage since the Devonian has increased the number of digits on its limbs. The giant panda alive today does have a sixth "digit" on its forepaws, but this is really a modified wrist bone, not an extra manus digit. All other examples of polydactyly are genetic phenomena that only affect certain individuals in a population, but are clearly not the species norm. Horses, a genus that normally possesses only one toe on each foot, are occasionally born with three toes, thus exhibiting a "throwback" to a more primitive condition.

As far as tracks are concerned, even in the case of 3-toed horses or 6- and 7-toed cats, the extra toes are almost always very small, or vestigial, and located in an elevated position on the side of the ankle or wrist, where they would not come into contact with the ground. Thus the chances of polydactyly showing up in the track record are very slight indeed.

One-of-a-Kind Tracks from the Eocene

In the western United States the track record of vertebrates that lived during the Eocene (58–37 million years ago) is not much better than for the Paleocene, in terms of published accounts. However, at least one formation is locally famous and shows considerable potential for further study. This is the Green River Formation, which is also famous for producing an abundance of fossil fish, plants, and insects, as well as rarer but well-preserved bones of birds and other vertebrates.

The Green River Formation is a source of oil shale and has been intensively studied by geologists interested in reconstructing the ancient environment. Such studies clearly show that the sediments accumulated in a series of long-lasting lake basins in what is now southern Wyoming, western Colorado, and eastern Utah (figure 6.5). Tracks, particularly those of birds, abound in certain shoreline sediments and have been extensively collected by both professionals and amateurs.

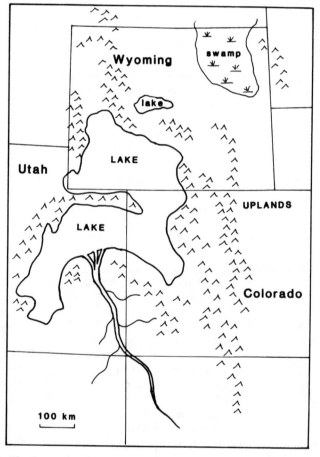

Figure 6.5 The location of Eocene Lake basins in relation to present-day geography.

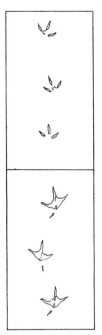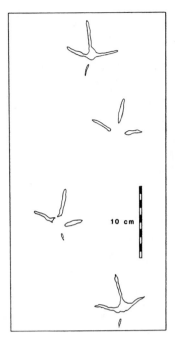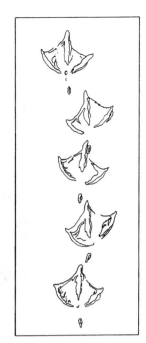

FIGURE 6.6 Common types of bird trackways from the Green River Formation. *Morphotypes after Greben and Lockley 1992.*

Preliminary studies of the tracks indicate at least four common and distinctive types of bird trackways from the Green River Formation (figure 6.6). These range in size from that of a small, plover-like species to that of a larger, web-footed species. The plover-like tracks are very abundant; the birds that made them left huge numbers of tracks along the lake shorelines. Until recently none of the four types of tracks had been given an ichnogenus name. Instead, two varieties of the small, plover-like tracks we simply labeled Type A and Type B. A larger wader we designated Type C (figure 6.7), and the web-footed species we have since named Type D.

According to paleontologist Alan Feduccia, the Type D trackway may have been made by *Presbyornis*, a bird that was transitional between flamingos and ducks. Accordingly, we formally named it *Presbyorniformipes feduccii*. Skeletal remains of the genus *Presbyornis* have been found in abundance in the Green River Formation. One particular trackway of this web-footed bird is famous for the associated bill or beak dabble marks, which indicate that it was feeding on organic matter in the lake shore substrate (figures 6.8 and 6.9). Sediments from the same formation have shown an abundance of small, sinuous trails (figure 6.7) made by nematode worms. Such traces are common along many modern and ancient lake shores and provide evidence of the abundant invertebrate life that sustained the shorebirds.

In addition to bird tracks, the Green River Formation has yielded trackways of amphibians, reptiles, and mammals. The class Amphibia, however, is represented by only one trackway—a single small frog that hopped across the mud flats (figure 6.10 and 6.11). Surprisingly, this is one of only two fossil trackways

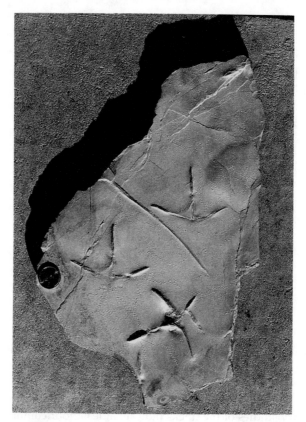

Figure 6.7 Bird tracks designated as Type C from the Green River Formation. Note abundant nematode trails covering the track-bearing surface.

made by frogs known from anywhere in the world. Reptile tracks from the Green River Formation are also scarce; only a single possible "lizard" trackway is known (figure 6.10). Mammal trackways are locally more abundant, as revealed by the spectacular "Strawberry Slab," a large specimen of track-bearing volcanic ash that was excavated in Strawberry Canyon in northeastern Utah. The tracks on this slab are all preserved as natural casts where the volcanic ash filled in original impressions in the muddy shoreline sediments. The specimen (figure 6.12) was reported by Shell Oil Company geologist H. D. Curry in 1960 and donated to the Smithsonian in Washington, D.C. This large slab shows the trackways of several three-toed mammals that were probably related to tapirs or ancestral horses.

Paleontologists have traditionally regarded the Tertiary history of the horse as a classic example of evolution. Its family tree can be traced through skeletal remains from the early Tertiary, when the Eocene "dawn horse," *Eohippus* (correctly known as *Hyracotherium*), was no bigger than a lap dog, down through various foal-sized intermediate forms, including *Miohippus* and *Pliohippus*, to the modern genus *Equus*. In western North America the skeletal remains of horses are quite abundant and well studied; however, little is known of their tracks. This is particularly unfortunate, as tracks could potentially reveal a lot about gait, behavior, ecology, and habitat of these various extinct forms—as we shall show shortly, in the case of another well-known group.

The trackmakers of the Strawberry Slab clearly overstepped their front footprints with their hind feet, a pattern that is typical of the walking gaits of modern mammals. The Strawberry Slab is one of the oldest examples of the trackways of odd-toed ungulates. It appears to provide evidence of several individuals moving in one direction—probably along a shoreline.

Although we know of no other large specimens of Eocene tracks that compare with the richness of the Strawberry Slab, at least two additional sites have yielded a few of these three-toed tapiroid or horse-like tracks. One site is also part of the Green River Formation; the other is in the Debeque Formation, which is about the same age. Like some of the footprints on the Strawberry Slab, one of these specimens shows evidence of the overprinting pattern that is typical of modern mammal trackways.

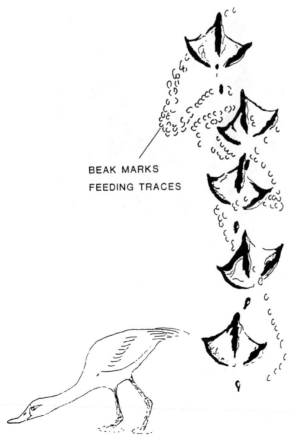

BEAK MARKS
FEEDING TRACES

FIGURE 6.8 Probable *Presbyornis* trackway, showing dabble marks or feeding traces. *After Erikson 1967.*

B.Y.U. B20

FIGURE 6.9 A probable *Presbyornis* trackway, *Presbyorniformipes feducii.* *Photograph of Brigham Young University specimen B.20.*

The Green River Formation and the Uinta Formation in this area have also yielded rare tracks of larger three-toed mammals. A good example of these large perissodactyl tracks is on display at the University of Utah Museum in Salt Lake City (figure 6.13). This particular slab also shows smaller three-toed mammal and bird tracks. At another locality in the Uinta beds of Wyoming, these typical tracks of a small three-toed mammal are found in association with bird footprints (figure 6.14).

Tracks have been reported from at least two other Eocene deposits in western North America: the Clarno Formation of Oregon and the Baca Formation of New

Mexico. The Clarno tracks were found by Bruce Hansen within the boundaries of the John Day Fossil Beds National Monument; these tracks provide evidence of a four-toed tetrapod, probably a small carnivore, that left cat-like tracks (figure 6.15). Although abundant skeletal remains of mammals and other tetrapods are known from the fossil beds overlying the Clarno Formation, this track discovery is the first for the Eocene of this area. The Baca Formation site in Socorro County, New Mexico, has yielded 18 footprints of camelid or, more probably, pecoran animals comprising three parallel trackways, and thus suggesting gregarious behavior. These trackways are among the oldest examples of cloven-hoofed tracks (figure 6.16).

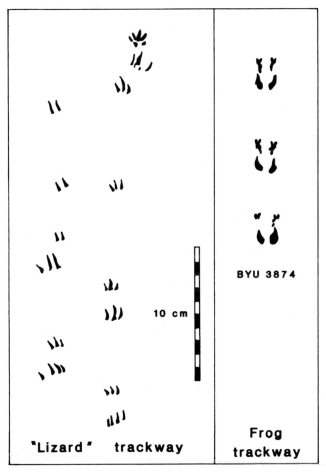

FIGURE 6.10 These two Eocene trackways from the Green River Formation depict a lizard (after Curry 1957) and a frog (after Stokes 1984). The frog trackway (see Figure 6.11 for a photograph of it) is the only Tertiary trackway of a frog reported anywhere in the world.

FIGURE 6.11 Trackway of an Eocene frog.

As if to prove that Eocene tracks are potentially as abundant as any others, recent reports demonstrate that a significant track record is coming to light. A spectacular large tracksite of the late Eocene has been reported in the Vieja Group of Texas. It was first studied, in 1975, as part of a University of Texas project. For two decades Bill Sarjeant and Wann Langston have been preparing a monograph that was just published in 1994. This substantial work describes the trackways of nineteen mammals, six birds, two turtles, and two invertebrates. This tracksite is without doubt the largest Tertiary tracksite yet discovered in the western states—if not the world. Not surprisingly, it shows a much greater diversity than anything hitherto recorded.

After having heard rumors about the forthcoming Sarjeant and Langston tracks monograph, we were thrilled to see it published just as this book was going to press. As we were lamenting the scarcity of useful documentation on Cenozoic tracks, this substantive study hit the bookshelves, proving what we already suspected—that Tertiary rocks are also full of tracks, just as they are in the Mesozoic and late Paleozoic. This timely account of west Texas footprints reveals a Trans-Pecos track bonanza dated as 36–38 million years ago (in the very latest Eocene). The tracks were made in volcanic ash in a situation reminiscent of the famous hominid tracksite in Africa, which is 3–4 million years old. Just as kin to Lucy left tracks in the Eastern Hemisphere on a carpet of ash, so our more distant mammalian relatives—including insectivores, carnivores, tapiroids, camelids, and others—left a rich track record in the Western Hemisphere.

The Puzzle of Miocene Tracks in the Oligocene

The track record of the Oligocene (37–25 million years ago) in the western United States is no better known than is the rather patchy Paleocene and Eocene records just described. But again, this impoverished track record is probably an artifact of limited fieldwork. This lack of attention is, however, changing.

Several Oligocene tracksites are now known in the western states. In 1943 Robert Chaffee described an assemblage of ungulate tracks from the White River Beds of Niobrara County, Wyoming. Chaffee noted that only two other specimens of Oligocene tracks were then known—one partial track housed in the Yale Peabody Museum and another in the American Museum of Natural History—and that neither had been recorded in the literature. Thus, he concluded that mammal tracks in the White River were "either rare" or had "received so little consideration" from collectors that few specimens had been recorded.

Chaffee was able to demonstrate that the Niobrara site yielded the tracks of both odd-toed, rhinoceros-like species and even-toed, camel-like species (figure 6.18). He also interpreted the partial Yale specimen as the front footprint of a brontothere, a herbivore and one of the largest land mammals known from the

FIGURE 6.12 The so-called Strawberry Slab in the Smithsonian Institution reveals some of the earliest known Eocene examples of odd-toed ungulate (perissodactyl) tracks—possibly made by ancestral horses.

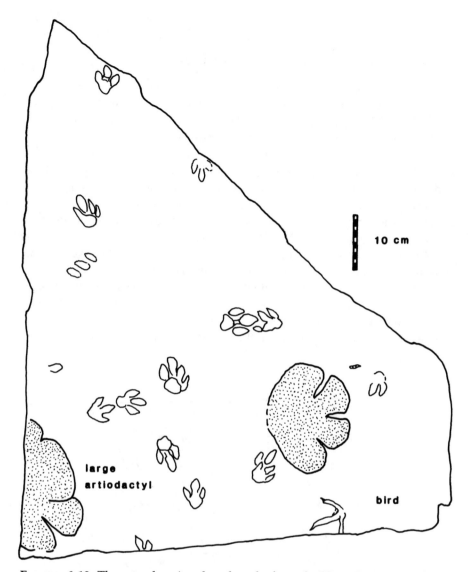

FIGURE 6.13 Three-toed perissodactyl tracks from the Uinta Formation of Eocene age. *University of Utah specimen.*

Tertiary. One of the most delightful aspects of Chaffee's report is the entertaining description of his efforts to obtain casts of modern rhino, tapir, and camel tracks at the Philadelphia Zoo. He found the Indian rhinoceros "Peggy" to be "reasonably tame" and willing to oblige "by making nice prints" of the front and hind feed in the wet clay and sand substrate prepared for her to walk on. However, "the tapir, being of uncertain temperament, presented a problem" and would not make tracks, so zoo staff had to hold the animal and press its feet into the substrate. Even this procedure proved difficult and the tapir "spoiled several attempts by kicking at the wrong moment." The camel was equally uncooperative and "seemed to want to hinder the progress of science" by refusing to walk on the prepared substrate even when led there by halter. Persistence did prevail, and a few tracks were eventually made and then cast in plaster (figure 6.18).

Chaffee described the Niobrara County sediments from which the tracks originated, noting that they represented "channel sand that filled in the original tracks" near the contact between the Oligocene White River beds and the overlying Pine Ridge beds of the Miocene. This observation is very interesting, not least because the tracks appear to have several characteristics suggesting affinities with Miocene animals. For example, the large size of the camel tracks suggested to Chaffee "the presence of a large camel hitherto unknown in the Upper Oligocene" and "twice as large as *Oxydactylus*, the largest mid-Tertiary camel." It is our opinion that the tracks Chaffee collected and described may well be those of Miocene animals. And yet, the original interpretation placed them in the uppermost beds of the Oligocene. Can we reconcile this apparent contradiction?

First, this situation is not unusual. We ourselves have seen Jurassic examples of tracksites at the contact between formations (see chapter 4 for discussion of tracks at the Wingate-Kayenta and Entrada-Summerville boundaries). In all these examples there is good reason to suspect that the tracks are associated with the younger, overlying beds rather than the older deposits. This is because tracks made in the underlying strata, before the accumulation of the overlying layers, stand little chance of being preserved for long at the surface. However, tracks may be made right around the time that sedimentation resumes—or just before, when the substrate is first wetted or washed over (figure 6.19). In any situation in which the break between lower and upper units represents a significant hiatus in

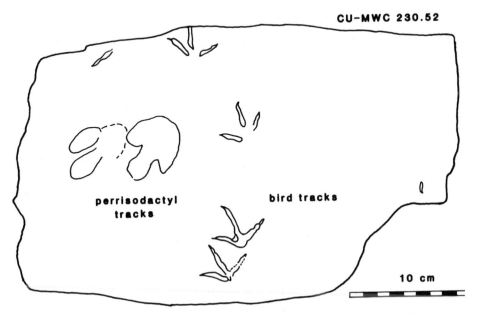

FIGURE 6.14 Tracks from the Uinta Formation, Eocene age.   *Source: courtesy of Ron Shult.*

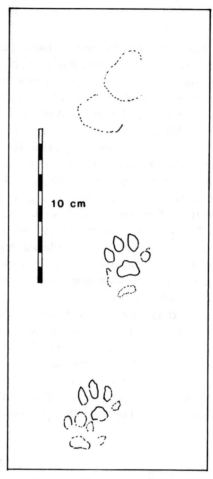

FIGURE 6.15 Eocene carnivore tracks from the Clarno Formation, John Day Fossil Beds. *Source: courtesy of Bruce Hansen and the National Park Service.*

time, the probability that tracks at the boundary correlate with the younger time, not older, becomes greater, and the importance of correctly dating the tracks becomes more critical. Conversely, the tracks might indicate that the original dating of the rocks is wrong; the boundary may well exist below the tracks rather than above.

Since Chaffee's study of the Niobrara County tracksite, at least seven track-sites have been reported from widely scattered Oligocene sites in the western United States that really are Oligocene in age. These include the late Oligocene Arikaree beds of Scotts Bluff National Monument in Nebraska, the Brule Formation of South Dakota, the Datil Formulation in New Mexico, the Mineta Formation of Arizona, and the Bear Valley Formation of Utah. Two of these sites reveal interesting track evidence worthy of further study.

In keeping with the current, unprecedented decade of track discovery, the White River Group has yielded an important new site. It was reported in 1993—the fiftieth anniversary of Chaffee's original report. Also proving the point that tracks are everywhere, this new footprint site was noticed in the middle of

Toadstool Park, a part of the Oglalla National Grassland reserve in Sioux County, Nebraska. Preliminary reports indicate that the track-bearing surface extends over a distance of one kilometer, and that it is part of an exhumed river valley. To date, at least eleven different vertebrate track types have been reported, including footprints of shorebirds, ducks, carnivores, rhinoceroses, camels, and other ungulates. Some of the bird tracks are associated with feeding traces, and some of the mammal tracks are found in groupings that suggest gregarious or social

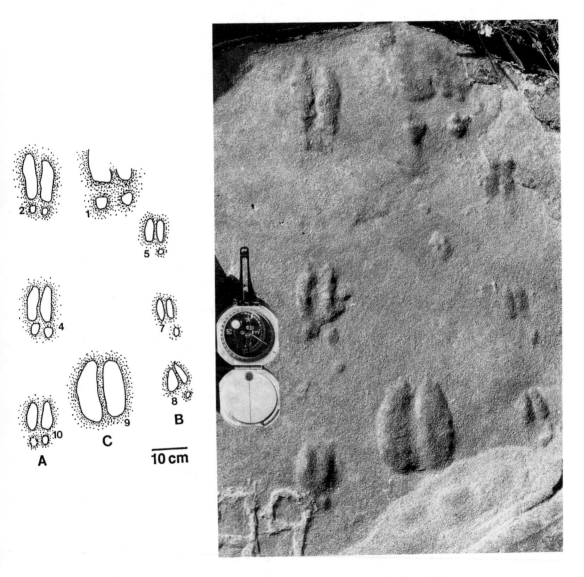

FIGURE 6.16 Parallel trackways of Eocene artiodactyl tracks from New Mexico may indicate gregarious behavior. *Source: courtesy of S. G. Lucas and A. P. Hunt.*

Big Bird Tracks Have Paleontologists All Aflutter

One of the most interesting tracks to be reported in recent years was the discovery of what is probably the first *Diatryma* track ever found (figure 6.17). John Patterson found it in Eocene sediments of the Puget Group at Flaming Geyser State Park in the western part of Washington State. *Diatryma* was a large, ground-dwelling bird that grew to a height of 2.3 meters and weighed in at close to 150 kilograms (about 350 pounds). It is sometimes depicted as a predatory bird, with powerful grasping claws and a large fearsome beak. The *Diatryma* featured in the film *Mysterious Island*, based on the Jules Verne classic of the same name, was depicted as a fearsome raptor. However, recent studies by Allison Andors suggest that it had hooflike "toe nails" better suited to walking than to perching or grasping. Andors also concluded that, rather than being a fleet-footed carnivore, *Diatryma* was a ponderous, slow-walking plant eater. This interpretation fits well with the plant-rich coal deposits of the Puget Group.

When Patterson announced his discovery of the single, 33-centimeter-long track, the *Seattle Times* ran a story, with this headline: " 'Big Bird' footprint has scientists aflutter" (3 May 1992). The story suggested that the track was a perfect match for a *Diatryma* footprint and that most experts, including Andors, concluded that the track was a genuine footprint of *Diatryma* or something similar. However, less than three months later, on July 17, the *Seattle Times* printed another story, this time titled, "Track is hoax, paleontologists say." This follow-up account described how, after close inspection, Andors and others concluded that the "track" might have been carved as a hoax. According to the newspaper article, the hoax, if that is what it is, was "expertly done" and there was no evidence that John Patterson had anything to do with it.

While we acknowledge that Andors is an expert on *Diatryma* and has had the opportunity to examine the track firsthand, it is our opinion that the track is probably genuine. We have examined several color photographs of the track and the site where it was found, and we have also talked to John Patterson and to officials associated with Flaming Geyser State Park. There are too many similarities between the alleged track and what a *Diatryma* track should look like for the resemblance to be coincidence. Could a practical joker understand the morphology of a *Diatryma* foot so well as to be able to carve a perfect fake that was the right size, with convincing hooflike toe impressions? Would a prankster, moreover, have chosen to do it in sedimentary strata of the right age and type? In our experience with tracks, carved hoaxes or forgeries are very rare, and they are never done with sufficient expertise to be convincing. Moreover, tracks are much more common than nontrackers believe, especially in coal beds.

Our conclusion, based on the available evidence, is that everyone's first impressions were probably correct. There is really nothing surprising about finding *Diatryma* or *Diatryma*-like tracks in Eocene sediments that represent the preferred habitat of this type of bird. As we have stated repeatedly in this chapter, the reason that more Tertiary tracks have not been found, until now, is not that they are rare, but simply that nobody has looked for them.

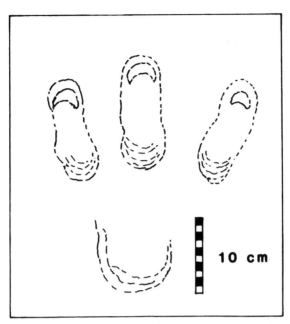

FigURE 6.17 Probable track of a *Diatryma* or *Diatryma*-like bird, from the Puget Group (Eocene) of Washington. This track is controversial, as it would represent the first such track ever found.
Source: drawn from a photograph by John Patterson.

behavior. In short, it appears that the tracksite is a window into the ecology of a riparian ecosystem adjacent to more open Oligocene prairie.

In a study of the Arikaree Group of Nebraska, David Loope claimed that he could recognize tracks in cross section at several different levels in sediments interpreted to represent a dunefield setting. He also concluded that the density of tracks was low probably because there were no large populations of animals in this environment, or because the sediments accumulated too quickly to allow time for any one surface to be extensively trampled. By contrast, tracks reported by Tony Fiorillo in the Mineta Formation of Arizona show high densities suggesting extensive trampling or milling around by large animals, possibly rhinoceros, in the vicinity of waterholes. As we discussed in earlier chapters, trampling is extensive in some late Mesozoic deposits where the tracks of large dinosaurs are common. But trampling has yet to be reported from early Tertiary deposits—presumably because large animals were rare, and because little attention has been devoted to searching for evidence of such stomping grounds. One should expect to find more examples of extensively trampled substrates in later Tertiary deposits, and this is evidently what has occurred in the Mineta sediments and in Miocene examples we discuss next.

Before we move to the Miocene, two other Oligocene sites deserve special mention. Artiodactyl tracks from predominantly eolian sandstone deposits have been discovered in the Datil Formation of the San Mateo Mountains near Socorro, New Mexico. There, tracks made in sand were filled in with volcanic ash fallout from a local eruption. The tracks are on display in the Mineral Museum at

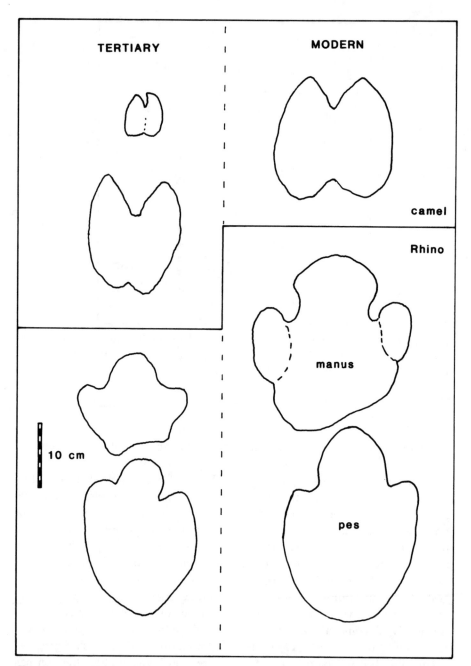

FIGURE 6.18 Rhino-like and camel-like tracks from the Oligocene(?) of Niobrara County, Wyoming. Modern tracks for comparison. *After Chaffee 1943.*

the New Mexico Bureau of Mines and Mineral Resources, Socorro. Another set of artiodactyl tracks was recovered from the Bear Valley Formation of eastern Utah. The specimen, which is in the University of Utah collections, was recovered from a siltstone layer above a volcanic horizon that has been dated at about 26 million years. This makes the tracks latest Oligocene to earliest Miocene in age.

Camels, Bear-Dogs, and Other Denizens of the Miocene

The Miocene epoch (24–5 million years ago) marks the beginning of Neogene time and the end of the Paleogene, which comprises the Paleocene, Eocene, and Oligocene epochs. Neogene tracks are generally better known and somewhat better documented than are the older tracks of the Tertiary. The trackmakers are often more precisely identified (i.e., at the genus level) in the later epochs of the Tertiary. This evidently reflects confidence derived from the fact that potential trackmakers are well known from skeletal fossils and are quite similar to modern animals. This accuracy of correlation is encouraging for the study of fossil footprints in general. As knowledge increases, the ability of paleontologists to match tracks with trackmakers will continue to improve.

Miocene tracks are known from scattered localities throughout the western states. But the tracksites in California—notably, in the Mojave Desert and Death Valley—are among the most informative. The best place to see examples of the California tracks is in the Raymond Alf Museum at the Webb School in Claremont, California.

Dr. Alf is responsible for the collection, documentation, and exhibition of several dozen Pliocene tracks from the Barstow Formation of the Mojave Desert. Among these are the tracks of birds, camels, deer, cats, elephants, and a large bear-dog. Several of these specimens are the centerpieces of truly spectacular displays in Alf's museum.

Raymond Alf, an indefatigable collector, went to considerable length to obtain segments of trackways for his displays. His emphasis on trackways—not lone tracks—means that the specimens often yield important data about step and stride, from which conjectures about locomotion and gaits can be derived. At one locality in the Mojave, he and his team of workers removed the overburden to reveal five camel trackways consisting of segments with between ten and twenty-six consecutive steps. The parallel configuration of these trackways (fig-

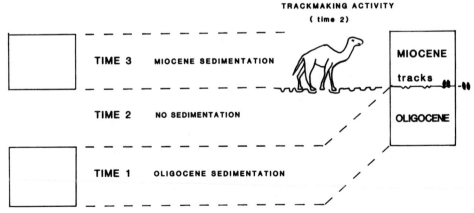

FIGURE 6.19 How Miocene tracks may have been made in Oligocene sediments. This is another example of why tracks tend to concentrate in the boundary zones between sedimentary units.

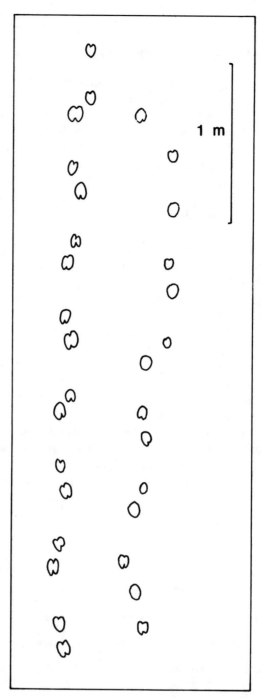

FIGURE 6.20 Miocene camel tracks probably made by *Procamelus* or *Protalabus* from the Barstow Formation, Mojave Desert, California. This specimen reveals two animals, side by side, moving with a pacing gait—which is very efficient in open terrain. It is on display at the Raymond Alf Museum in Claremont, California.

ure 6.20) suggests that the animals were traveling as a group. Alf and one of his students, David Webb, suggested that the trackmakers were probably *Procamelus* or *Protolabus*. The trackways are long enough and well enough preserved, in their original trackway configurations, to allow for a fascinating study of Miocene camel gaits.

Ancient Ships of the Desert

One of the classic stories in Tertiary paleontology is the evolution of the prairies in North America. This evolution was driven by a progressive cooling and drying that led to the decline of woodlands and the emergence of grasslands, steppes, and prairies. The shift was also evident in the Eastern Hemisphere, with the evolution of the African savanna. These changes, which were most pronounced in late Oligocene to Miocene time, had a profound effect on the flora—which meant, in turn, that animals had to adapt to eating tougher, drier vegetation and to living in open country.

Thus many of the mammals became larger, with anatomical changes (notably, hooves) that made it possible for them to move swiftly across the wide, open spaces. This is the traditional, if perhaps rather simplified, scenario of how the horse grew to be big and fleet of foot. This, too, accounts for buffalo, deer, and antelope evolving into the indigenous plains fauna we know so well.

But perhaps the best example of the evolution of hoofed animals is that of the camel—a creature we don't immediately associate with the Western Hemisphere. In Tertiary time, however, the camel (like the horse) experienced a spectacular evolutionary radiation—disappearing from North America (again, like the horse) only very recently. Unlike the horse, the camel was never successfully reintroduced.

In an elegant and compelling article about the evolution of camel locomotion, published in 1972, David Webb revealed how the track record is important in tracking these ancient ships of the desert. (David Webb is a former student of Raymond Alf's; Alf inspired half a dozen of his high school students to go on to become accomplished professional paleontologists.) Webb was able to demonstrate from his study of camel trackways and anatomy that camels were one of the few groups of large, hoofed mammals to develop the pacing gait. Pacing, as compared with trotting, involves simultaneously moving both limbs on the same side rather than diagonal pairs. This pacing mode of locomotion allows for a longer stride, and a lesser expenditure of energy to cover an equal distance. The modern camel is the only wild animal today that regularly employs this gait. Webb's work reveals that early in the camel lineage the pacing gait was already well developed. In developing this pacing gait, camels became very long-limbed; they sacrificed maneuverability for efficiency. Webb believes that the camel's pacing ability played a major role in its conquest of the desert, where efficient modes of travel between patchy, and far-removed food and water sources would have been advantageous.

(continued)

The track record, like the bone record, shows that camels became very abundant about Miocene time. Moreover, their success at this time coincides with the spread of arid conditions and arid floras into the region that is now the southwestern United States. Despite our images of buffalo, deer, and antelope at home on the range, camels were in some ways the epitome of adaptation to Tertiary prairie and drier habitats, with their own unique gait. They were perhaps the first prairie schooners!

Alf also collected a beautifully preserved segment of the trackway of a large carnivore, probably the bear-dog *Amphicyon*. Known from skeletal evidence to have been a giant member of the bear-dog family, *Amphicyon* had hind footprints 28 centimeters long and a 1.3-meter stride. It probably stood 1.6 meters at the shoulder and weighed about 250 kilograms. This "dog" was thus about the size of an adult grizzly bear. Alf was able to extract five consecutive footprints (figure 6.21) by cutting through the rock with a skill saw powered by a portable generator. He then used a banding machine to bind the pieces for transportation to his museum. The effort was successful; the reassembled trackway is now a stunning display at his museum. The American Museum of Natural History in New York has recently replicated the trackway and mounted an *Amphicyon* skeleton above it for a dynamic new exhibit.

The Barstow Formation consists of a complex sequence of mudstones, sandstones, limestones, and volcanic ashes that have been dated, using the potassium-argon technique, at 15–16 million years old. The sediments represent lake, river, and ash deposits that accumulated in landlocked, low areas in "basin and range" terrain typical of that part of the western United States then and still somewhat today. (Great Salt Lake is a prime modern example of a land-locked basin.) The sediments and fauna of the Barstow Formation indicate more humid conditions than those that prevail in this arid region at the present time. The track assemblages—with bird, camel, gazelle or deer, dog, and cat—are characteristic of many track assemblages from Neogene deposits of similar age.

A younger Miocene track-bearing deposit in the Mojave is the Avawatz Formation, dated at about 11–12 million years ago. Like the Barstow beds, the Avawatz has also been prospected and documented by Raymond Alf. It reveals an assemblage of bird, camel, dog, and cat tracks. About twenty such specimens are on display in the Alf museum.

Probably the best-studied Miocene tracksite in the western United States is in the Copper Canyon Formation of Death Valley National Monument. Recent work by Paul Scrivner and David Bottjer of the University of Southern California has given insight into the animal communities that lived in this area and the environments they inhabited 7–10 million years ago. Again, the track assemblage is quite familiar, consisting of tracks of birds, camels, deer, horses, carnivores (cats and bear-dogs), and elephants. About a half dozen representative tracks are

FIGURE 6.21 Trackway of a large bear-dog, probably *Amphicyon*. *After Alf 1966.*

on display in the Alf Museum, including a fine specimen with horse and camel tracks on a surface with raindrop impressions.

The data presented by Scrivner and Bottjer in their 1986 report were drawn from over a hundred track specimens. This sample is large enough to shed light on the composition of the animal community (figure 6.22). The track census indicates a community dominated by camels (possibly *Procamelus*) and llama-like trackmakers. Bird and horse tracks were also relatively common, but the tracks of deer, cats, bear-dogs, and elephants are uncommon.

Scrivner and Bottjer presented some interesting interpretations of the Death Valley environment of the late Miocene. They clearly demonstrate that the track-bearing beds represent lakeshore sediments, but they also showed that lake levels rose and fell as climatic conditions fluctuated between wetter and drier. There is evidence that tracks are more abundant during wet or rainy (pluvial) periods when water was fresh and substrates wetter—thus attracting animals and creating good conditions for track formation. By contrast, footprints were less common during arid phases when evaporation caused the lake systems to shrink, substrates to be drier, and conditions unfavorable for animals and for track formation. The presence of assemblages of tracks is an indication of reasonably favorable habitats in the region. But as Scrivner and Bottjer point out, many of the trackmakers may have been just passing through. Track density and the rarity of heavily trampled zones suggests that the (likely salty) playa lake—

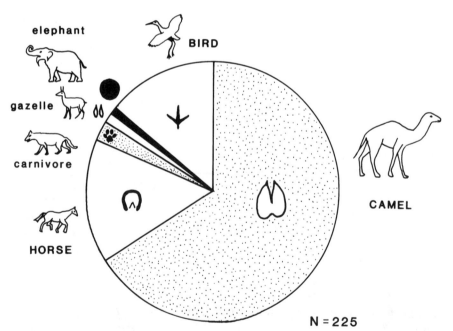

FIGURE 6.22 The Miocene animal community from Death Valley, based on track census data drawn from 225 occurrences of tracks or trackways. Camels left far more trackways than did any other kind of trackmaker. *After Scrivner and Bottjer 1986.*

especially during dry phases—probably lacked much in the way of resident animal populations. However, these authors did include an illustration of one beautiful example of a trampled layer, thus showing that there was significant activity at certain times.

During late Tertiary time, the present-day basin and range physiographic province had been taking shape, owing to what geologists call crustal extension (figure 6.23). The basins form natural areas for lakes to develop. Even today when the climate is arid, scattered alkaline and ephemeral lakes are widespread throughout the basin and range province. In the Neogene, when rainfall levels were generally higher, these basins often contained much more water. A record of these lakes is found in the shoreline sediments, which can also reveal a lot about the climatic and physiographic history of this tectonically active region. However, geologists do not agree on the best way to identify shoreline sediments. That is where ichnologists may have something special to offer. Experience has shown us that tracks are sensitive indicators of shores or strandlines. We therefore predict that increased knowledge of the distribution of track beds in the west will lead to an improved understanding of the history of lake levels in this area, which will in turn improve the ability to interpret climatic fluctuations and tectonic activity.

Although the best-known Miocene tracks have so far been recovered from California, several interesting sites occur in other western states—notably, Nebraska and New Mexico. To date, however, there is little documentation of these sites. All we know is that camel tracks have been reported from the Ash Hollow Formation of Nebraska and that camel and carnivore tracks have been observed in the Ogallalla Formation (Miocene-to-Pliocene age) of New Mexico. Bird tracks are also known from the Tesuque Formation of New Mexico, near Española. Further exploration and study of such sites promises to improve the overall knowledge of Miocene footprints in the western United States.

The Kansas extension of the Ogallalla Formation offers one further ichnological clue about the Miocene. In a 1986 publication, Joseph Thomasson, Michael Nelson, and Richard Zokrzewski reported the preservation of fragments of grass in Ogallalla Formation deposits in Kansas. These grass fragments appeared to have been trampled into the sediments after deposition. The fragments revealed excellent preservation, which allowed the authors to determine the type of photosynthesis employed by the grasses. Microscopic features indicated C4 photosynthesis, which is generally uncommon among plants, but known to occur in grasses adapted to high temperatures and light intensities. Thus, these researchers obtained compelling evidence that C4 photosynthesis was established at least 5 to 7 million years ago.

It is hard to prove that trampling was directly responsible for the successful burial and preservation of the grass fragments. Would the fragments *not* have been preserved in the absence of trampling? In our view, the vertebrate bioturbation may well have been a contributing factor. It is widely accepted that trampling of vegetation and of invertebrates—for example, the record of sauropods trampling clams, discussed in chapter 4—can induce the burial and preservation of organisms that would not otherwise have become fossilized.

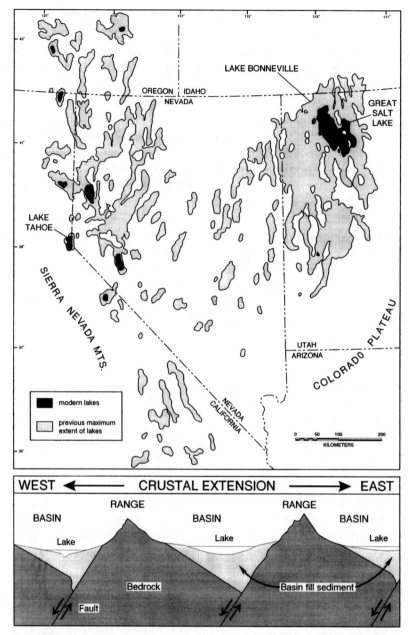

FIGURE 6.23 Geographic extent of Pleistocene-age lakes, compared with modern lakes. Cross-section shows typical basin and range country and the resulting development of lakes.

Pliocene Tracks and the Camel That Died While Standing

The Pliocene epoch began only 5 million years ago and ended 1.6 million years ago at the onset of the great surges of glaciation that characterized the Pleistocene. Alas, Pliocene tracks are no better known or documented than those from older Tertiary deposits. However, a few sites are known and hold promise of yielding additional material. These sites occur in three rock formations: the Muddy Creek Formation of southern Nevada, the Tecopa Formation of the Mojave Desert area of California, and the Bidahochi Formation of Arizona.

The Nevada locality has yielded mammal tracks (camels and carnivores) along with bird footprints. The bird tracks by themselves are noteworthy for their variety and range of size (figure 6.24). The largest indicate the presence of very large heron-like waders. They are by far the largest *undisputed* bird tracks recorded up to this point in the Tertiary, and they compare in size with the tracks of living Goliath Herons and Maribou Storks of Africa. Examples are housed in the University of Utah and the University of Colorado.

The Mojave Desert specimens from the Tecopa Formation include camel and carnivore tracks that are, however, not fully documented. These specimens are on display at the Raymond Alf Museum in Claremont, California. At one locality in the Tecopa beds, paleontologists found remains of camel limbs in an upright position, indicating that an animal had become mired and had died while still standing. They nicknamed the locality "standing camel basin." Such examples of animals that were mired or stuck in their tracks are not unique. Leaving aside obvious examples of death traps like the La Brea Tar Pits, there are reports of other Tertiary mammals that got bogged down and even of Jurassic dinosaurs that apparently suffered similar fates.

Finally, Bidahochi tracks from Arizona include camel footprints and the tracks of two types of bird. The type with web impressions closely resembles the track of the modern Canada goose (genus *Branta*). The other type represents a large, heron-like wader. Two track slabs are preserved in the Museum of Northern Arizona collections. The track-bearing beds also contain casts of ice

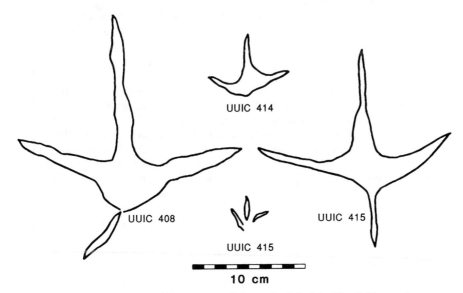

FIGURE 6.24 Bird tracks (Pliocene age) from the Muddy Creek Formation, near Glendale, Nevada, show a variety of forms and sizes.

crystals, showing that freezing periodically affected Pliocene lake environments in Arizona, though not necessarily at exactly the same time as the tracks were made. Given the wide range of daily and annual temperature fluctuation in Arizona at the present time, this evidence of freezing is not unexpected.

Mammoths of the Pleistocene

The Pleistocene epoch (1.6 million to 12,000 years ago) marks the beginning of the Quaternary period. The Tertiary is over, and the climate has shifted. The Great Ice Age is another name for the Pleistocene epoch.

During the Pleistocene, large ice sheets accumulated over the northern parts of North America, Europe, and Asia. These continental ice sheets were accompanied by smaller ice caps and glaciers that developed in high altitude regions, like the Rocky Mountains and the Alps. The history of the ice sheets is complex and varied. The cold, glacial periods, when ice advanced, alternated with warmer and wetter times of glacial retreat. The combination of ice cover and erosion by ice created conditions in which tracks could neither be made nor preserved in northern latitudes. South of the ice, however, wetter and warmer conditions prevailed as a result of meltwater and the filling of basins with lakes that were generally much more extensive than at the present time. It is around such lakes and other wet substrates that the Pleistocene track record accumulated.

Linking Tracks with the Trackmakers

Frank Peabody, who was later to make his name in paleontology by completing extensive and exemplary studies on Triassic tracks, began his career as a tracker, in 1940, by studying Lower Pliocene tracks from a locality in the flanks of the Sierra Nevada in east-central California. Peabody compared the Pliocene trackways of amphibians with those of "relatively unchanged" modern descendants found along the Pacific coast and in the Appalachian mountains.

In an unrelated study, Peabody reported that he was able to differentiate trackways of living salamanders down to the genus level. He found reliable distinctions in the marks left by *Triturus*, *Ensatina*, *Aneides*, and *Batrachoseps*, while studying their trackways on mudflats along the San Francisco Peninsula. Out of this study of living trackmakers, Peabody arrived at a remarkable conclusion: a trackway of a tetrapod is, as a rule, generically distinctive. This is a very significant claim because it suggests that trackers will ultimately be able to distinguish between most fossil trackways at the low taxonomic level of genus. And why not? After all, today's trackers specializing in living organisms can generally identify the tracks to the genus level (if not further). And hunters living in intact indigenous cultures would likely be equally capable in this regard. Paleontologists studying fossil footprints can therefore hope to make significant progress toward this ultimate goal.

Among the sparse documentation of Pleistocene tracks from the western United States are reports of bear tracks from Oregon, camel tracks from New Mexico, mammoth tracks from South Dakota, miscellaneous mammal tracks from California and Arizona, and a rich assemblage of bird and mammal tracks from Nevada.

Bear tracks thought to be of Pleistocene age were reported from Lake County, Oregon, in 1980. These footprints are up to 40 cm long—making the trackmaker as big as any bear alive today—and the claw impressions are distinct (figure 6.25). The footprints are part of a trackway nine meters long. The trackmaker had a step of about 1.25 meters. The hind foot overstepped the forefoot in walking. This gait is typical of bears, and it has been suggested that the trackmaker may have been *Arctotherium*—the only known Plio-Pleistocene mammal capable of

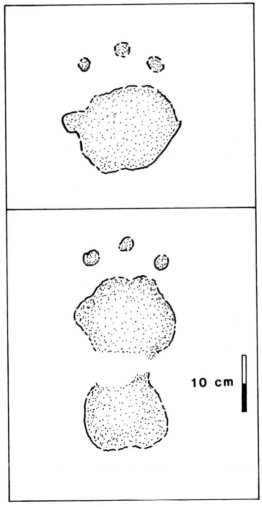

FIGURE 6.25 Two separate bear tracks from Plio-Pleistocene deposits of Oregon. *After Packard and Allison 1980.*

making such prints. This genus of bear is known from the famous mid-Pleistocene tar pits of Rancho La Brea in the Los Angeles area.

As explained previously, camel or camel-like tracks are common in Neogene deposits of the western states, as our brief review of Miocene and Pliocene sites has already indicated. European invaders of North America found no camels, but camels were still present into the Pleistocene. There are reports of camel tracks from the Popotosa Formation of the Santa Fe Group in Valencia County, New Mexico. At this site a continuous trackway of a relatively large animal (foot length 19 cm) has been preserved in sandstones associated with an alluvial fan deposit.

The mammoth is probably the best known of all Ice Age mammals, and many discoveries have shown that it was a common animal. Because of large size its remains did not easily disintegrate, thus making bone fossils relatively easy to find. It follows also that large mammoth tracks should be relatively easy to find, and, as we shall discuss, this proves to be the case.

A mammoth tracksite in Hot Springs, South Dakota, is important because it is so unusual. Within the city limits of Hot Springs, this one site has yielded the bones of over thirty subadult and mature mammoth individuals. It is a spectacular sinkhole deposit of late Pleistocene age (at least 20–25 thousand years old). The depositional environment was not the lake shoreline typical of other tracksites, and a lot of bones were buried along with the tracks.

The sinkhole evidently developed when a cylindrical area of bedrock collapsed, following the dissolution of underlying strata. The resulting breccia-choked pipe filled with artesian water (figure 6.26), creating an oval pond no more than 50 meters long by 30 meters wide. This sinkhole pond acted as a trap for sediment—and as a death trap for mammoth. As the sediment accumulated, mammoth left their tracks at various levels in the pond deposits. It is particularly noticeable that the tracks appear sporadically in the lower and middle parts of

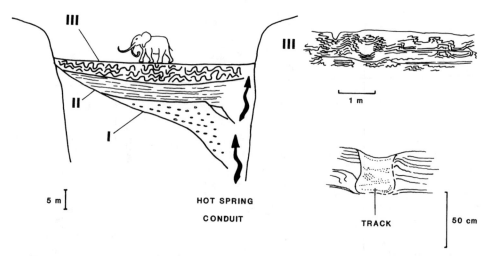

FIGURE 6.26 Mammoth tracks at Hot Springs, South Dakota, help reveal the story of a mammoth death trap. Note intense trampling (megabioturbation) in upper unit III; details shown to right.   *After Laury 1980; Agenbroad 1984.*

the sequence and become abundant toward the top, where the sediment is thoroughly trampled. The trampling evidence is so spectacular that geologist Robert Laury coined the term "megabioturbation" to describe it.

The many layers of tracks demonstrate that mammoth came to the sinkhole repeatedly over a period of time. The bones show that some became trapped after slipping or sliding into the steep-sided pond, from which it was hard to escape. Initially, however, the animals did not completely trample the sediment. Later, as water depth was reduced by increased accumulation of sediment and decreased spring discharge, evidence of trampling is more widespread. Increased trampling suggests perhaps a reduction in the rate of sediment accumulation, or it could indicate increased activity by the animals that visited the site. In either case, many bones within the sediment show evidence of fracturing that resulted from trampling. Mammoth were indeed capable of causing considerable damage when they trod on recently buried bone. The trampling evidence supports the scenario of shallow water at the end of the pond's history by showing that the trackmakers were exerting full force on the substrate rather than being buoyed up by water. Thus, tracks at the Hot Springs site are an integral part of the evidence for reconstructing the history of this unique mammoth graveyard.

Another mammoth tracksite has been reported from a locality known as "Elephant Hill" in the vicinity of Montezuma Castle National Monument in Arizona. The tracks occur in limestone of the Verde Formation and include those of camels and carnivores, along with the mammoth footprints. Tracks have evidently been reported from at least two localities in this area. A collection of footprints was purchased by the American Museum of Natural History. According to published illustrations, the mammoth tracks are 40–45 cm in diameter, with a single stride of about 2 meters. Although the trackway superficially appears to be that of a bipedal animal (figure 6.27), the trackmaker was obviously a quadruped that overstepped its front footprints with its hind feet, as is done by modern elephants.

The Case of the Carson City "Man Tracks"

It is a Pleistocene trackway that gave rise to one of the most entertaining episodes in the history of paleontology. This is the story of the Carson City tracks. These tracks were evidently made on the shores of a lake some 50,000 years ago, when Nevada enjoyed a more humid climate than it does at present.

In the early 1880s, at the Nevada State Prison, inmates were quarrying sandstone when they uncovered a remarkable series of over fifty fossil trackways. What is more, these trackways were diverse. They included varieties easily attributed to mammoth, deer, wolf, and birds. But the highlight was about ten trackways that were at first attributed to a giant human wearing sandals.

In 1986 Jordan Marche looked back on the episode and concluded that these particular tracks are among "the most interesting, yet least known . . . in the history of American paleontology and rightly deserve greater recognition than they have so far achieved." We second his assessment. The tracks are important on

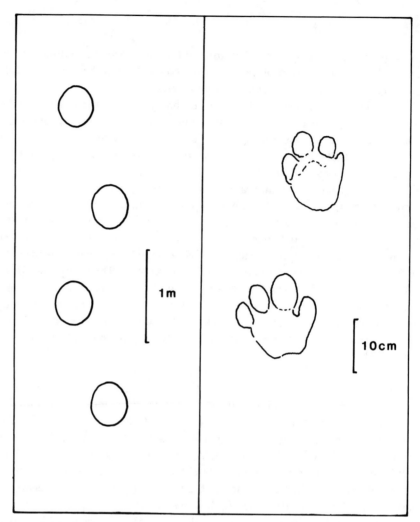

FIGURE 6.27 Trackways from the Plio-Pleistocene Verde Formation at "Elephant Hill" in Arizona. *Left:* mammoth trackway (after Brady and Seff 1959). *Right:* a pair of carnivore tracks (after Nininger 1941).

two counts: first from the perspective as a chapter in the cultural history of science and second for their intrinsic scientific value.

As discussed by Marche, during the 1860s and 1870s, following publication of *The Origin of Species* and *The Descent of Man* by Charles Darwin, the American public was easily "stirred into high-pitched frenzies" by the announcement of discoveries of various petrified human remains. Many of these later proved to be frauds or hoaxes—like the famous Cardiff Giant of New York, which turned out to be a statue that had been buried deliberately. This was also a time, however, when genuine remains of human ancestors were being unearthed in both Europe and North America. The problem was that archaeologists and geologists at this time had a very poor understanding of the age of most of these remains; radiometric dating techniques were yet to be invented. And knowledge about human evolution during the Pleistocene epoch was virtually nonexistent.

The practice of paleontology was further confused and clouded by various fraudulent activities and the nonscientific articles that inevitably followed. One such contribution was penned by no less a literary giant than Mark Twain. In 1862, in an amusing attempt to satirize the high-pitched frenzy of ancestry hunting, Twain wrote an article for a newspaper in Virginia City, Nevada. In this article, titled "The Petrified Man," Twain set out to destroy the "growing evil" of what he called "this ridiculous mania." But, upon publication, he found that "nobody ever perceived the satire," and his yarn was taken as truth. In 1884 Twain wrote another paleontological spoof—this one on "The Carson Fossil Footprints." He attributed the human-looking tracks in Carson City, Nevada, to the drunken staggering of "primeval" members of the Nevada Territorial Legislature on a rainy night.

But what were the Carson City tracks, if not those of humans? The earliest report—by William Harkness—had correctly identified the mammoth, deer, wolf, and bird tracks but had attributed the sloth tracks to "imprints of the sandaled foot of man." This interpretation was hastily supported by the great paleontologist Edward Cope. A lesser-known individual, the University of California geologist Joseph Le Conte, however, disagreed. We now know that Le Conte's was the correct interpretation.

Le Conte attributed the tracks to that of a giant ground sloth, for which no tracks had yet been discovered (figure 6.28). Cope's arch rival, Othniel Marsh, sided with Le Conte and showed how the extinct ground sloth *Mylodon* possessed a foot skeleton that exactly fitted into the tracks. W. P. Blake also supported the sloth interpretation. Moreover, he judged that the site as a whole, in all its grand diversity, was the most important fossil footprint discovery "since the unearthing of the fossil (dinosaur) footprints in the sandstones of the Connecticut Valley."

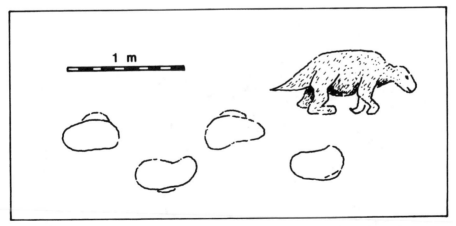

FIGURE 6.28 Tracks of a giant ground sloth, probably *Mylodon*, from the Pleistocene of Nevada. These tracks were first interpreted as those of a giant human wearing sandals.   *Source: based on a replica on display at the University of California Museum of Paleontology, Berkeley.*

Although the tracks at Carson City were covered up by construction at the prison in the late 1930s, a few specimens had been removed and are still on display at the Nevada State Museum in Carson City. Additional specimens are housed at the Mackay School of Mines in Reno and at the University of California at Berkeley. The most useful record is probably a map of a big chunk of the site, produced by Addison Coffin in 1889 (figure 6.29). This map shows the exact configuration of dozens of trackways. And it confirms the diversity of Ice Age animals active in this area.

As shown in the census diagram (figure 6.30), the most common trackways are those of a sloth (probably *Mylodon*), followed by heron-like bird trackways and those of mammoth, deer and elk, horses, and finally carnivores (dogs and

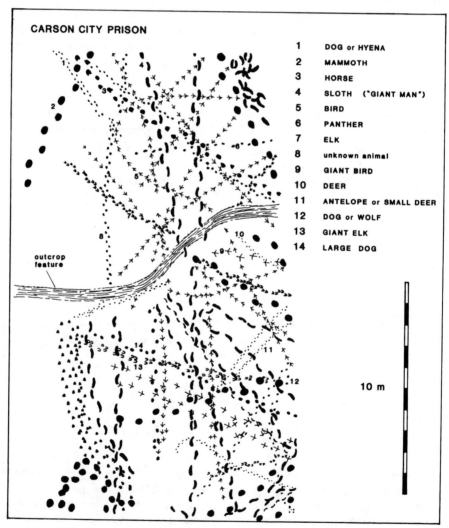

1	DOG or HYENA
2	MAMMOTH
3	HORSE
4	SLOTH ("GIANT MAN")
5	BIRD
6	PANTHER
7	ELK
8	unknown animal
9	GIANT BIRD
10	DEER
11	ANTELOPE or SMALL DEER
12	DOG or WOLF
13	GIANT ELK
14	LARGE DOG

CARSON CITY PRISON

outcrop feature

10 m

FIGURE 6.29 Map of the Carson City Prison tracksite.   *After Addison Coffin.*

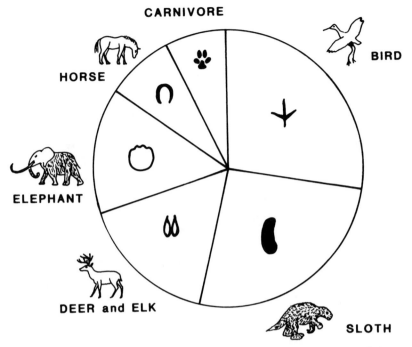

FIGURE 6.30 Census of trackmakers derived from Carson City tracksite map, based on a total of 54 trackways.

cats). The overall herbivore/carnivore ratio among the mammals is 35:4, or approximately nine to one. This is in reasonably good agreement with predator/prey ratios in modern mammal communities. Thus, we can visualize an animal community with typical modern elements like deer, dogs, and cats—but also the giant sloth, mammoth, and prehistoric horse.

Cruising on the Dinosaur Freeway. A herd of small quadrupedal ornithopods heads north while large ornithopods head south. Based on Cretaceous tracks from the Mosquero Creek site, New Mexico.

7

TRACKS GALORE, AND WHAT
THEY CAN TELL US

*Vertebrate ichnologists have succeeded in demon-
strating how important their discipline is to
understanding both the paleobiological and environmental significance of vertebrates.*

GEORGE PEMBERTON (1993)

■ Part of the motivation for writing this book was to come to grips
with the large amount of track information that is now available in the western
United States and to try to understand what it can tell us. Given that the scientific study of tracks had been neglected until quite recently, it has proved relatively easy for us to discover new sites and to revisit or restudy old ones. To date,
we have recorded more than 500 of these new and old tracksites. We and our
research group colleagues discovered a large number of them just in the last
decade. Such a large source of information has already led to a growth and
maturing of the science of vertebrate ichnology, or tracking, to such an extent
that results from this field are having an increasing impact on the field of ichnology as a whole and on the field of vertebrate paleontology (figure 7.1).

Paleontology, like other branches of the earth and life sciences, has a strong
descriptive component, so we devoted much of this book just to describing the
tracksites—old and new—and the site-specific information that could be
gleaned from them. We have also discussed the significance of the track evidence
in shedding light on various aspects of the biology of ancient trackmakers and
the worlds they inhabited. Now, in our concluding chapter, we consider the
broader implications of this landslide of new footprint data that the western
United States has to offer.

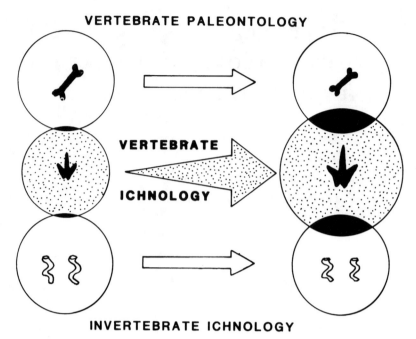

VERTEBRATE PALEONTOLOGY

VERTEBRATE

ICHNOLOGY

INVERTEBRATE ICHNOLOGY

FIGURE 7.1 Schematic representation of the growth of the field of vertebrate ichnology and its importance relative to other disciplines.

Naming Tracks and Identifying Trackmakers

Ichnologists, including trackers, recognize that fossil footprints may not always reveal the identity of trackmakers with certainty. But tracks are often interesting in their own right. They offer insights about animal behavior, ecology, and the ancient environment. Nevertheless, trackmaker identification is and will continue to be a prime objective for all trackers. It is therefore worth reviewing how much progress has been made in trackmaker identification.

In the past, tracking has been considered a dubious pastime, because there was so much uncertainty about "who dun it." Such uncertainty still exists, but the situation is much improved. Whereas, in the past, trackers could not always key a track or trackway to even the major classes of animals (amphibians, reptiles, birds, and mammals), it is now possible in many cases to distinguish trackmakers down to the level of order, family, or even genus. Thus, confusion about the appearance of footprints made by amphibians, mammal-like reptiles, crocodiles, pterosaurs, turtles, and other groups is largely or partially resolved.

Some trackmakers, including the late Cretaceous king of beasts, *Tyrannosaurus* (probably *T. rex*), along with the late Triassic aquatic reptile *Tanytrachelos* and certain late Tertiary and Quaternary mammals, have been identified, with reasonable confidence, down to the level of genus or species. The ability to make these accurate identifications depends to a great degree on a process of elimination that involves looking not just at track shapes and trackway patterns but also at the age of the strata in which the tracks occur and the skeletal remains, if any, known from deposits of that age. Thus, for example, *T. rex* is the only known animal that could have made the track depicted in figures 5.44 and 5.45 at the time the track was made. Any claims of *Tyrannosaurus* tracks from earlier or later epochs, regardless of track shape, can be ruled out on the basis of age.

Overall, trackmaker identification is ideally a three-pronged approach that involves consideration of track shape, track age, and the body fossil record. Furthermore, every time trackers narrow down the identity of a trackmaker, the overall track record becomes more precise, and more useful for vertebrate paleontology as a whole.

Paleontologists, including trackers, tend to assume that the track record is sketchy and incomplete whereas the bone record is good and relatively complete. This assumption is only true in some senses, and in the case of some formations. Obviously, a complete skeleton provides more information than does a trackway, but a significant number of species of extinct animals are defined on the basis of only a few bones that make up only a small fraction of a complete skeleton. In such cases a good trackway may provide as much information as do a few bones, and so it is just as legitimate a part of the fossil record as are the skeletal scraps. Moreover, while trackers sometimes appear to be playing "catch up" or trying to match the track record with the skeletal record, there is just as much need for those who study bones to try to match their skeletal data with the footprint record. The latter, however, is something that has seldom been done. In the future, therefore, we hope that paleontologists will use this book, and other track literature, to try and find the footprints that correspond to their skeletal discoveries.

Finally, on the subject of trackmaker identification, we note that the increase in discovery of new tracksites inevitably leads to an increase in the number of well-preserved trackways available for study. This in turn allows ichnologists to distinguish well-preserved tracks from those that are poorly preserved and to begin the slow process of reevaluating past work in order to determine invalid or duplicate track names. As this process of reevaluation unfolds, the names that are applied to tracks will become increasingly more precise and meaningful. Ultimately, the effort will lead to the suppression of invalid or dubious names, and it will perhaps result, finally, in a lesser number of well-defined names— rather than the large number of poorly defined names that today makes comparative ichnological work difficult and confusing.

Individual Behavior

During the course of the twelve years of field work and specimen analysis that made this book possible, we had cause to ponder a number of instances in which specimens revealed interesting behavior of the trackmaker. Why, for example, are certain unusual trackway patterns found on the lee or avalanche slopes of sand dunes? These trackways (see figures 2.5 and 2.13) are unusual in that they zig-zag and change gait patterns. Could such trackways reflect running behavior of mammal-like reptiles? If so, *when* did tetrapods first develop the ability to run, and can we discover from track evidence which groups were the first runners, and how fast they ran? The track record has the potential to answer these questions and to shed light on important questions about the evolution of locomotion and the contribution of improved gaits to the success of certain groups.

Similarly, specific track specimens encourage us to ponder the late Jurassic theropod (an allosaur perhaps; figure 4.45) and the early Cretaceous ornithopod (an iguanodontid; figures 5.21 and 5.22) each of which had suffered injuries that left them limping. Track evidence will probably never tell us how these animals sustained their injuries, or how severe they were, but we already suspect that the injuries were not life threatening and that the iguanodontid, at least, still kept up with the herd. Because we now know of at least a half dozen other examples of limping dinosaurs from around the world, we can also use the track record to tell us the frequency of injury and which groups of animals were afflicted. To date, such injuries appear to have been rare, but both carnivores and herbivores sustained injury.

We can also ponder the behavior of the flying pterosaurs, long considered to be rather cumbersome creatures on land. Some researchers have argued that pterosaurs were bipedal, like birds. The trackways show, however, that they actually walked on all fours, leaning forward with much of their weight supported by their front knuckles. Trackways also indicate that many pterosaurs congregated on marine mudflats; surely they took advantage of onshore breezes to launch themselves from there into Mesozoic skies.

Groups, Populations, and Social Behavior

Our survey of tracks in the western United States has led us to conclude that there are indeed "tracks galore." The sheer abundance of tracks offers great utility in assessing the relative populations of trackmakers in certain environments and at certain times. Mammal-like reptile tracks are abundant in desert dune environments, but scarce elsewhere. *Brachychirotherium* (aetosaur?) tracks are common in the latest Triassic deposits; theropod tracks are common in the early Jurassic. Hadrosaur tracks are very common in late Cretaceous deposits, whereas *T. rex* footprints are exceedingly rare.

Overall, we conclude that by studying enough tracks it becomes possible to distinguish the common from the rare. And sooner or later the investigator is tempted to look further, into patterns of distribution that invite analysis and interpretation.

Notably, there is increasing evidence that many of the large herbivorous dinosaurs were gregarious. Not only do large numbers of tracks of sauropods and ornithopods occur at many late Mesozoic sites, but a significant number of sites also show evidence of herding, in the form of large numbers of parallel trackways. For example, during the course of this study we documented the Mosquero Creek site (figure 5.20), where more than fifty individuals appear to have been traveling north. This is the largest herd of dinosaurs reported on the basis of track evidence from North America.

Now that tracks and other lines of evidence have helped establish that many types of dinosaurs were gregarious, we can ask what the evidence suggests for other groups. Were the prosauropods, the predecessors of the brontosaurs, gregarious or not? Sufficient track evidence is now accumulating to suggest that the preliminary answer may be yes, at least in some cases. What about the archosaur (aetosaur?) responsible for making *Brachychirotherium* tracks? And was the mammal-like reptile that left *Laoporus* tracks gregarious? When did gregarious behavior become established among large tetrapods? among which groups? and can we demonstrate such behavior from the track record? These are all intriguing questions, and we have no reason to doubt that, in the not too distant future, tracks will help answer such questions, either fully or in part.

As a direct result of studying large assemblages of tracks, we have learned that tracks often reveal interesting data on population structure. For example, some herds consist mainly of small individuals, others of large individuals, and yet others of a mixture of sizes (figure 7.2). Within a particular community or population, thus, animal herds were segregated by size. This presumably means that they were also segregated by age, or possibly by sex. Recent work on the relationship between dinosaur size and age, based on skeletal studies, suggests that ichnologists, too, can begin to assess the age structure of herds, based on their own data. Such an approach has been used in Africa by wildlife biologists studying elephant populations, and there is every reason to expect that the same principles can be applied to dinosaurs and other extinct animals—providing fossil data permit reasonable estimates of age.

Dinosaur Biomass

In order to understand aspects of the social behavior and the population and community ecology of dinosaurs and other extinct creatures, it is important to develop a substantial numerical database. In this way we can establish whether animals were rare or abundant, and we can learn about the relative abundance of large and small animals within particular species or groups. As emphasized repeatedly throughout this book, the trackway database is very large in numerical terms, and, as just shown, useful for estimating the sizes of individuals in various track samples or "populations."

Paleontologists working with bones rather than tracks also count individual animals and record their sizes. If both number and size of individuals is recorded, then it is possible to estimate biomass. However, estimating biomass is controversial for two reasons. First, studies of modern ecosystems show that we can expect to find that small animals are more common than large ones. Unfortunately however, the fossil record generally distorts numbers by preserving too few remains of small animals, and an overrepresentation of large animals. As discussed in chapter 4, this problem is probably more acute for bones than for tracks, though there is also some degree of bias toward preserving large tracks. Second, estimates of biomass, especially between predator and prey species, are important in understanding both modern and ancient community dynamics. In the case of the fossil record, for example, it has been claimed that, among dinosaurs, predator:prey biomass ratios of about 1:30 indicate that dinosaurs had large food requirements, like modern mammals, and so were warm-blooded. If the ratio was only 1:3 it would indicate much smaller food

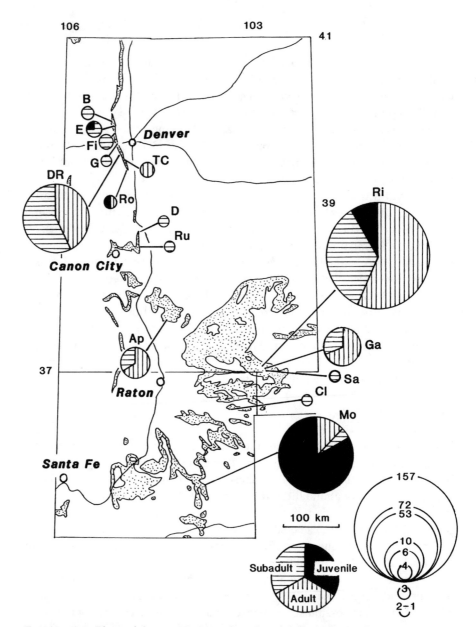

FIGURE 7.2 Plots of the proportion of tracks of different sizes from various outcrops of the Dinosaur Freeway (Dakota Group of Colorado, Oklahoma, and New Mexico). Keys at bottom for age of trackmaker and sample size (the biggest circle is used for the site with 157 trackways). Note the large proportion of juveniles from Mosquero Creek (Mo) compared with Richardson ranch (Ri) and Dinosaur Ridge (DR). *Abbreviations for site names after Lockley et al. 1992.*

requirements on the part of the predators. (Note also that, any uncertainty about which species were predators and which were prey, introduces uncertainty into the discussion).

Because tracks are so abundant, they are potentially useful for shedding light on biomass, community dynamics and predator:prey ratios. For example, the tracks of theropods or carnivorous dinosaurs (predators) are often much more abundant than those of herbivores (prey species); see chapter 4. It has been suggested that this is in part due to the high activity levels of theropods. Until now however, nobody has ever thought to convert numbers of tracks into biomass estimates. To show the value of such an exercise we estimate the biomass of theropods and sauropods from the famous Jurassic Purgatoire site in Colorado, and show how the results compare with simple numbers. A count of trackways shows that there are about 60 theropod trackways and 40 sauropod trackways associated with the main track level (figure 4.49). Such a large number of theropod trackways, is exactly the sort of evidence that raises questions about the track record. Many paleontologists rightly question why they appear to be so over-represented, when in the bone record they only make up 3–5%, instead of 60%.

If we estimate biomass however, we arrive at a very different conclusion. Based on average theropod track size (foot length of 37 cm), at the Purgatoire site, we can estimate that the average animal weighed about one ton. By contrast, the average sauropod had a foot length of about 67 cms. Because volume (and mass) increases as the cube of the linear dimension, it is soon obvious that the sauropods had a much greater biomass. Our estimates for the average animal are in the range of 9–10 tons. Thus the 60:40 ratio becomes a 60:360, or 60:400 ratio. As shown in figure 7.3, this produces a very different picture of the predator:prey ratio. The predator biomass is only 15–17% of the total, not 60%. Thus we can argue that the results obtained using this method are probably more meaningful because they are closer to those obtained from bone census data.

In future we hope that we will be able to produce reliable estimates of biomass of different trackmaking groups from the abundant track census data that we are beginning to compile. Such an approach will require some standardization of how trackers estimate size and weight (biomass) of individual trackmakers. But, as the previous example shows, the results are promising in terms of arriving at biomass estimates that more accurately reflect ancient ecology and community dynamics than numbers alone.

Trampling

When the first tetrapod walked on land sometime in the Devonian, the only other animals responsible at that time for disturbing soils and substrates were invertebrates, such as molluscs, worms, and arthropods. These animals burrow and feed in soils and substrates, disturbing sedimentary layers and leaving telltale traces referred to as "bioturbation" or "ichnofabrics." At first, vertebrates caused little bioturbation and can hardly be viewed as destructively trampling the environments they inhabited. Throughout the late Paleozoic, however, as vertebrates expanded in terrestrial habitats, they had an increased impact on substrates. By Permian and Triassic time some of the first heavily trampled areas were preserved in the rock record, providing evidence that vertebrates were locally abundant, or at least repeatedly active in certain areas.

This phenomenon of vertebrate trampling or disturbance of substrates increased dramatically in the late Mesozoic, when vertebrates, especially sauropod and ornithopod dinosaurs, became very large, abundant, and gregarious. The result was a new phenomenon—not just local vertebrate trampling or bioturbation, but widespread "dinoturbation." The sedimentary rock record clearly shows that by the end of the Mesozoic large dinosaurs were trampling extensive areas and having a significant impact on substrates over much wider areas. Large herds of large mammals that can be "tracked" today provide a living analog of just how destructive bigness combined with abundance can be. Large herds of herbivores throughout the Mesozoic and Cenozoic, whatever their taxonomic affiliations, must have created pockets of devastation quite regularly.

One leading dinosaur expert, David Norman, has suggested that trampling acted as a stimulus to plant evolution, causing fast-growing or rapidly regenerating species to have a distinct advantage over slow-growing varieties. This trampling and browsing pressure may, according to Norman and others, have been a factor contributing to the rise of flowering plants, which are fast-growing and which have dominated the land environment since the end of the Mesozoic era.

Our own work indicates that it is possible to measure the level of trampling or dinoturbation by simply plotting the number of sites, over time, where trampling has been recorded. Figure 7.4 shows that the frequency of dinoturbation increased throughout the Mesozoic. This means that geologists can expect to find trampled layers in strata of this age, particularly of the late Mesozoic, that are more abundant and more pronounced than almost anything else encountered in the rock record before or since. The only exception is to be found in rocks of the very late Cenozoic, which yield evidence of a good deal of trampling by elephants and other large mammals. The trampling pattern through time depicted in figure 7.4 is a remarkable depiction of the impact of dinosaurs throughout the late Mesozoic and the abrupt curtailment of trampling that accompanied the end-Cretaceous extinction.

Tracks and Ancient Ecology

An abundance of tracks is useful not only for investigating social behavior but also for developing a census of ancient animal communities. Moreover, because tracks are usually more common than bones, they reveal evidence of many more individuals than can usually be obtained from the skeletal record. In addition, physical evidence is almost always easier to obtain for tracks, and tracks represent the living—not the dead.

Our recent three-year survey of tracks in several formations at Dinosaur National Monument yielded evidence of more than 250 individual trackways during the latest Triassic and early Jurassic. This sample is larger than all the individuals recorded from skeletal evidence quarried from the Monument (though from a different time horizon) over a period of about eighty-five years. Similarly,

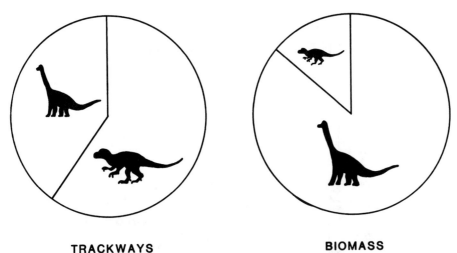

TRACKWAYS   **BIOMASS**

FIGURE 7.3 Biomass estimates (right) are probably closer to the true predator-prey ratio than trackway counts. (Based on the Late Jurassic Purgatoire site, Colorado.)

our census of the trackmakers recorded in late Triassic Chinle deposits within the Dinosaur National Monument area is based on over a hundred trackways (figure 3.24). Census data are available, or can be compiled, for many other track-bearing deposits of all ages, which we will continue to pursue. For example, the large track-site at Purgatoire River in Colorado records approximately 100 individuals living at the same geologic "moment" in the late Jurassic, from which a census can be compiled. Megatracksites, such as the Cretaceous "Dinosaur Freeway" of the Dakota Group of the Colorado Front Range and western high plains, offer even larger samples for analysis. In a recent compilation we reported about 200 well-documented trackways from the Dakota Group, including about 50 from Dinosaur Ridge. From this we compiled the census depicted in figure 7.5. Further work we conducted at Dinosaur Ridge, Mosquero Creek, and elsewhere has substantially added to this data base, bringing the total of well-documented trackways in the Dakota Group to more than 300. This figure does not include the trampled surfaces in which individual trackways cannot be discerned, and it does not include a number of other trackways that are known, yet not adequately documented.

As noted in earlier chapters, such data can be used in a number of ways. The data can be used to show the number of trackways of a particular type at a particular site, or combined to show the total for several tracksites in a particular sedimentary deposit. The data could also be used to show the geographic distribution of particular trackway types in a particular deposit, as shown in figure 7.6, where the distribution of ornithopod, theropod, crocodilian, and bird tracks is plotted for the Dakota Group study area.

In our recent study of the late Cretaceous Laramie Formation near Denver, where over 60 trackways have been recorded (figures 5.40 and 5.41), we have used the data to plot the relative abundance of trackmakers and the frequency with which they occur at different sites (figure 7.7). But we have also calculated how the proportions change through time from one track-bearing level to the next

Frequency of
TRAMPLED SUBSTRATES

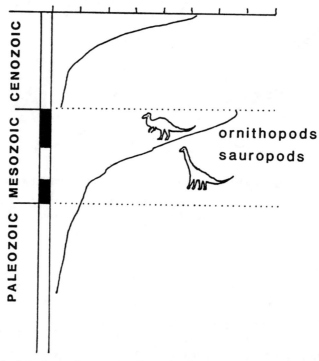

FIGURE 7.4 A plot of vertebrate trampling through time reveals an increase in the abundance of trampled sites in both the late Mesozoic and (following the mass extinction) again in the late Cenozoic. The body fossil record does indicate that vertebrates were evolving into larger forms at these times. The late Mesozoic peak of dinoturbation is mainly attributable to the activity of large gregarious sauropods and ornithopods.

(figure 7.8). As most of this data have only recently been compiled, its utility for paleoecological analysis has yet to be fully explored. And in addition to providing basic census data through time, track data can also provide information on the relative numbers of small (juvenile) and large (adult) individuals of any given type, as well as information on direction of travel, and so on.

The skeptic might well charge that it is risky to derive ecological and other conclusions from the percentages of track types derived from a census of a few tracksites. How indeed do we know that data retrieved from such sites are usefully representative of animal communities that once existed in the area? Our answer is that if tracksites of the same age and region or habitat repeat the same census patterns, then ecological conclusions are probably justified. Repeatedly we do find, as in the case of the Dakota and Laramie examples just cited, that track assemblages, sometimes called *ichnocoenoses*, or simply *ichnofaunas*, show recurrent patterns in a particular sedimentary facies. This by definition constitutes an *ichnofacies*, which is a recurring record of an animal community and the habitat that supported it.

Paleontologists studying invertebrate fossils in marine and aquatic environments and ichnologists studying invertebrate traces have long known that recurring patterns are meaningful. As a result, biofacies and invertebrate ichnofacies have been defined and related as recording the same ancient ecosystems (that is, both the ancient animal community and the environment that it inhabited). In like manner, we and other ichnologists believe it is possible to recognize and define bona fide vertebrate ichnofacies and to know that they are not random assemblages of tracks.

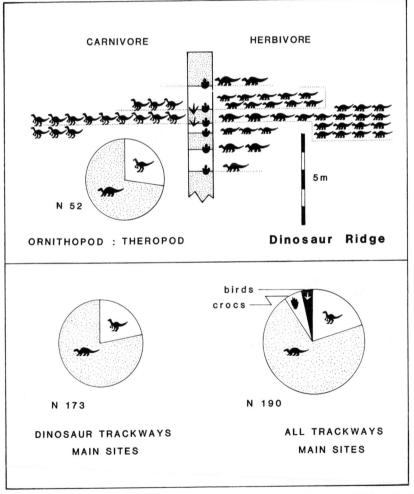

FIGURE 7.5 Census of vertebrates based on tracks from Dinosaur Ridge (top) and from the entire Dakota Group (bottom). Sample sizes are noted by N. *After Lockley et al. 1992.*

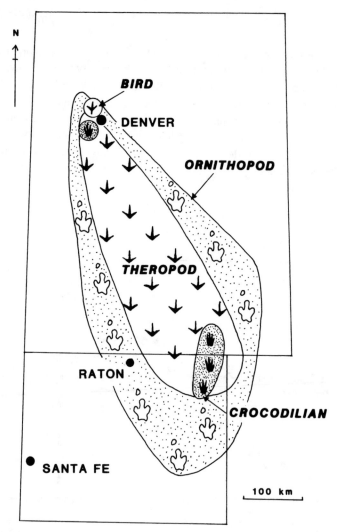

FIGURE 7.6 Geographic distribution of track types within the Dakota Group megatracksite area. Ornithropod tracks are found throughout area.

Sedimentological Controls on the Distribution of Track-Bearing Strata

Recognition of vertebrate ichnofacies underscores the fact that tracks are an integral part of the sedimentary rock record. Thus it is to be expected that the distribution of track-bearing layers will be controlled to a significant degree by sedimentological processes. In the same way that bones may be concentrated in river channel deposits or deposits where sedimentation has been rapid, tracks are usually concentrated along shorelines or at sites where the substrate coincides with the water table. While it is outside the scope of a book on footprints to discuss these broad patterns of distribution of track-bearing sediment, some observations are pertinent.

First, tracks are concentrated in certain types of sedimentary deposits. Most important are lake strandline sediments, river overbank deposits (such as crevasse splay sediments that result from periodic flooding), and even desert playa and interdune deposits that are subject to sporadic inundation. These are the *local* environments where one can expect to find tracks.

What about much more extensive track-bearing deposits, such as megatracksites? At least two types of regionally extensive track-bearing deposits are known. One, like the Entrada-Summerville transition zone of the mid to late Jurassic, is confined to a very thin, but laterally extensive, zone, which could be characterized as essentially a single surface or bed. Examples include the Glen Rose Formation of Texas of Cretaceous time, as well as a newly discovered upper Jurassic megatracksite in Switzerland. In these multiple-layered sites, the layers can be correlated with changes in sea level, and the tracks appear to be associated primarily with the beginning of sea-level rises—or the onset of transgressions.

The second type of megatracksite is a significantly thicker, but nevertheless laterally extensive, track zone. The track-rich zone in sequence 3 of the Chinle Group and in sequence 3 of the Dakota Group (figure 5.11) fall into this category, as does the track-rich Cretaceous Jindong Formation of South Korea. The Dakota Group is also associated with the onset of a transgression, but the Jindong and the Chinle are associated with continental basin systems (figure 7.9) rather than with coastal plain deposits, and so they must be understood in terms of the dynamics of the evolution of sedimentary basins.

The challenge facing ichnologists is to understand exactly what mechanisms lead to the formation and preservation of these megatracksites and track-rich zones. If we understand the sedimentological controls on the formation of such deposits, we can hope to use this knowledge to predict where to find more tracks.

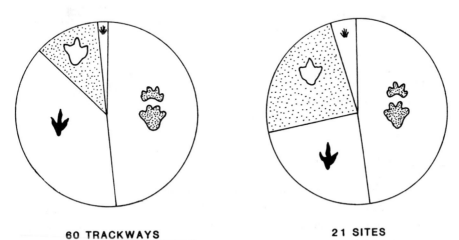

60 TRACKWAYS   **21 SITES**

FIGURE 7.7 Census of track types from the late Cretaceous Laramie Formation near Denver, based on total number of trackways (left) and frequency of occurrence of track types at tracksites (right). Track symbols refer, clockwise from top, to ceratopsians, theropods, ornithopods, and champsosaurs.

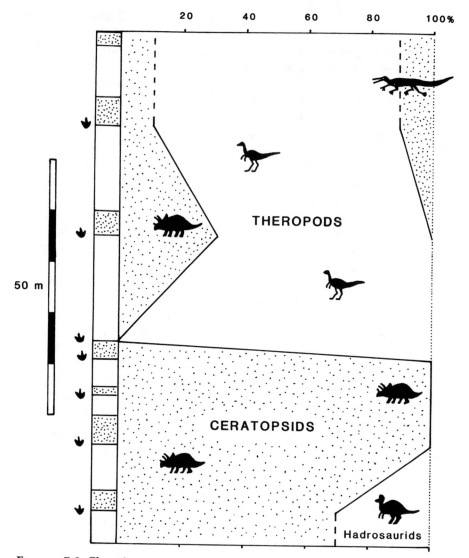

FIGURE 7.8 Changing proportion of track types at successive track levels in the Laramie Formation. Notice the decrease in hadrosaur tracks and the increase in theropod tracks through time.

Tracks More Common Than Bones

Consideration of the factors that lead to the preservation of track-rich deposits ultimately leads to the broader question of the completeness of the entire track record, and how it compares with the skeletal record. A perennial question in paleontology revolves around the completeness of the fossil record. Traditionally, the track record has been considered both sparser than, and inferior to, the bone record in terms of both quality and quantity. Such a viewpoint, derived largely from ignorance of the extent of the fossil footprint record, is ripe for reevaluation.

Vertebrate paleontologists who focus on the study of skeletal remains may consider the track record inferior to the skeletal record in terms of quality. This is largely because most of the reasonably well known fossils have been identified to the species level, whereas few if any tracks can be proven to belong to a trackmaker identified down to the level of species. In paleontology, after all, bones provide the type specimens by which organisms are known and by which reconstructions of the flesh-and-blood beasts can be made. The genus and species names used for tracks (ichnogenera and ichnospecies) fall under an entirely separate taxonomic scheme. The independence of ichnotaxonomy is therefore essential for preventing hasty and incorrect correlations with putative trackmakers.

Few trackers would dispute the notion that bones are the best way to get to know an extinct vertebrate, especially in the case of deposits that have good bone records. But tracks are much more common than bones. And there is a virtue in sheer abundance. Consider: each individual animal has only one skeleton, but it can potentially make thousands of tracks. One dinosaur may leave its tracks over a lot of acreage during its life time, and even in a range of depositional environments. But a dinosaur leaves its skeleton in only one spot—and that spot may not be a good one for achieving immortality. Moreover, the forces of nature may carry the corpse to an environment that the animal would never have visited when active. Bones, thus, are rare, and they cannot be trusted to always tell us how and where the dinosaur lived. Nevertheless, not until today's renaissance in the field of vertebrate tracking have trackers been able to demonstrate that abundance truly does count for something. Tracks can indeed tell us a lot of interesting things that bones simply cannot.

The relative abundance of tracks and bones can be determined in at least two ways. First, one can take a single formation and count the bones and tracks within it. But a lot of bone-rich formations apparently have few tracks, and likewise some track-rich formations lack much in the way of bones. Although bones and tracks do occur together much more often than received wisdom would have it, the bone-track ratio varies from formation to formation, depending on a variety of factors. Therefore, analysis of a single formation may not tell us much about the big picture of track and bone abundance.

A better approach is to make a list of all formations in a particular study area where data are available, and then to classify each formation according to the relative abundance of bones and tracks. Table 7.1 shows the results of such an effort we made for our study area: the western United States, and covering Triassic, Jurassic, and Cretaceous time. When we use this scheme to classify all the reasonably well documented Mesozoic deposits—which represent about half of the entire track record discussed in this book—we find that track-rich deposits significantly outnumber bone rich-deposits (figure 7.10). (To avoid bias in the sample, we included the well-known, bone-rich deposits from the study area, and still came out with a preponderance of track-rich deposits).

What do these results mean?

If bones and tracks were distributed equally in the fossil record, then every formation would fall into the category we list in table 7.1 as Type 3. A Type 3 category is one in which footprints and skeletal remains are equally abundant.

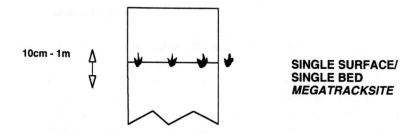

10cm - 1m

**SINGLE SURFACE/
SINGLE BED
*MEGATRACKSITE***

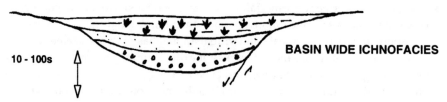

10 - 100s

BASIN WIDE ICHNOFACIES

FIGURE 7.9 Geometry of large, track-bearing deposits or megatracksites reveals two categories: extensive single surfaces and track-rich facies with multiple layers.

Geologists know, however, that this is not the case and that different types of sedimentary deposits have different potential for preservation of different types of fossils, depending on physical and chemical conditions in the environment—such as, acidity, current energy, and sedimentation rate. Given these differences in the distribution of track or bone fossils in different types of sedimentary deposits, it is clear that one must census all (or a randomly selected subset of all) the fossil-rich deposits and then classify each as to relatively more track-rich or bone-rich. This we have done. The results depicted in table 7.1 bring us to the conclusion that track-rich formations are more common than those in which bones predominate.

We are thus faced with the implication that, from a purely quantitative perspective, paleontology has overestimated the importance of bones and underestimated the importance of tracks—or at least overestimated the importance of the body fossil record relative to the track record. Overall, tracks are an abundant, yet underutilized, source of information. And we suggest that the track and bone records are *both* integral components of the whole picture. As such, they should both be fully utilized to establish as complete a data base as possible.

One key benefit that is sure to come from a more balanced use of bone and track data in paleontological interpretations is that it will help researchers to better recognize trustworthy from untrustworthy data. Notably, if the two records are consistent, the data are likely good. If the two are not in accord, data from either set should be used more cautiously because the track and bone records are obviously sampling different components of the extinct vertebrate community. In the latter case, moreover, the only way to obtain a reliable picture of the past is for both sources of data to be put to use. And tracks will finally be given their due.

A Finite and Valuable Resource

Although tracks are more abundant than bones, they, too, are a finite and nonre-newable resource. As pointed out by paleontologist Dan Chure, of Dinosaur National Monument, a species cannot get any more endangered than already being extinct. Some fossil species are known from only a single fossil, and if that fossil is stolen, destroyed, or lost, for any reason, the species can essentially become extinct for a second time.

In recent years, fossils of all kinds have escalated in monetary value consid-erably, thus encouraging an already lucrative commercial market. While we are

TABLE 7.1　*Classification of Terrestrial Track- and Bone-bearing Deposits*

Stratigraphic Unit	Ichnofacies	Track v. Bone Record Category
Cretaceous:		
Hell Creek Fm.	—	*4*
Two Medicine Fm.	—	*4*
Judith River Fm.	—	*4*
Raton Fm.	ornithopod	*2b*
Laramie Fm.	ceratopsian*	*3a*
Mesaverde Gp.	ornithopod	*2a*
Dakota seq. 3	*Caririchnium**	*1*
Dakota Gp. (other)	—	*3b*
Glen Rose Fm.	*Brontopodus**	*2b*
Cedar Mountain Fm.	—	*4*
Jurassic:		
Morrison Fm.	—	*3b*
Summerville Fm.	—	*2b*
Entrada Fm.	—	*2b*
Carmel Fm.	—	*1*
Navajo Fm.	*Brasilichnium** (in part)	*2b*
Moenave / Kayenta Fms.	—	*2b*
Wingate Fm.	—	*1*
Triassic:		
Sheep Pen Fm.	—	*1*
Chinle Gp. seq. 3	—	*2b*
Chinle Gp. pre-seq. 3	—	*4*
Moenkopi Fm.	—	*2b*

NOTE: Classification of deposits as to relative abundance of tracks and bones is in accordance with the scheme of Lockley 1989, 1991:

　1 tracks only
　2 tracks > bones
　3 tracks = bones
　4 bones > tracks
　5 bones only
　a track and bone records largely consistent
　b significant inconsistencies in the track and bone records.

* vertebrate ichnofacies previously defined (Lockley et al. 1994).

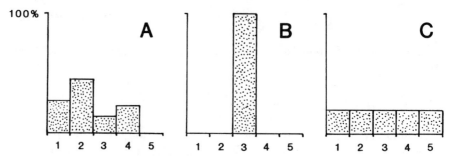

FIGURE 7.10 Plot (A) of the number of track-rich and bone-rich facies in the Mesozoic of the western United States, based on data in table 7.1. Categories on the x axis range from deposits that contain only tracks (1) to those that contain only bones (5). It is clear that track-rich deposits predominate over those that are bone-rich. For comparison, plot B depicts the hypothetical situation in which all deposits have equal numbers of bones and tracks. Plot C depicts the hypothetical situation in which all five categories of bone- or track-rich deposits were equally represented in the fossil record.

not in a position to judge the merits of commercial trading of fossils, much of which takes place within the broad realm of museum science and education, we do note that several unfortunate incidents of track theft have taken place on public lands. Outright theft and other acts of vandalism have diminished the scientific value of a number of important sites in the western United States. Even in the case of sites on private land that are exploited legally, we see much interesting material on the market that has not been given the scientific attention that it deserves. Moreover, in the case of all fossils, but for tracks especially, when they are removed from the ground without adequate documentation, valuable information is inevitably lost.

We conclude therefore that, despite the abundance of tracks in the western United States, fossil footprints are valuable and should be preserved for future generations of scientists and museum visitors. Like any other fossil resource, tracks are vulnerable to destruction by weather and erosion, and difficult decisions may be involved in how to preserve tracksites, especially large sites, in river valleys, or other high erosion areas.

As much as we trackers love tracks, we unfortunately have no special skills or resources to make track beds resistant to the relentless forces of erosion. We cannot always convince local, state, or federal authorities to institute costly measures to preserve particular sites. And such efforts must ultimately be done on a site-by-site basis, as necessity or community pressure dictates. However, we do believe that the first step toward preservation is documentation. That means the establishment of a permanent record through scientific publication, the making of replicas, and the repositing of valuable and vulnerable specimens in museums. This book is our contribution to this essential task.

APPENDIX

FOSSIL FOOTPRINT SITES AND EXHIBITS IN THE WESTERN UNITED STATES

Arizona

LAKE POWELL

Glen Canyon National Recreation Area, Arizona and Utah
Mainly Triassic and Jurassic periods
Dinosaur tracks

Lake Powell is the reservoir created by the massive Glen Canyon Dam. A number of fossil footprint sites have recently been discovered along the lake shore and in adjacent areas, from the southern end near Page, Arizona, to the northern end near Hite, Utah. Large theropod tracks (*Eubrontes*) removed from Explorer Canyon are on display at the Visitor Center, at Page, and other tracksites in this area are featured in some of the glossy guidebooks available to tourists (Barnes, 1986; Davies, 1992). One of the most spectacular sites is an outcrop of tracks on an overhang, about 150 feet above lake level, at Tapestry Wall, north of Bullfrog. Kelsey's 1989 guide to Lake Powell also provides directions to a tracksite in the Rincon area. Here visitors can see theropod tracks in the Wingate Formation if they are prepared to hike about a mile from the lake shore.

Museum of Northern Arizona

Flagstaff, Arizona
Mainly Permian, Triassic, and Jurassic periods
Tracks of predinosaurian reptiles and early dinosaurs

The Museum of Northern Arizona in Flagstaff has a substantial collection of Permian fossil footprints from the Coconino Sandstone of the Grand Canyon Region. As with all collections from this sandstone, the predominant track type is *Laoporus*, attributed to mammal-like reptiles. There are also *Chirotherium* tracks from the Triassic Moenkopi Formation and dinosaur tracks from the lower Jurassic Glen Canyon Group. Flagstaff is the main turnoff point for travelers on Interstate 40 who wish to travel north to the Grand Canyon, Lake Powell, or to tracksite areas at Tuba City or further north in eastern Utah.

Tuba City

Navajo tribal lands in Northeastern Arizona
Jurassic era
Dinosaur tracks mainly attributed to theropods (carnivores)

A locally well known dinosaur tracksite is situated on Navajo Nation tribal lands, just west of the small town of Tuba City on Highway 160 between Cameron and Kayenta at the turn-off for the Moenave. The tracks are of two kinds of lower Jurassic theropod, assigned to the ichnogenera *Kayentapus* and *Dilophosuripus* (see chapter 4). Roadside signs are in place to direct visitors to the site. About a mile east of the turn-off a new site has been discovered and signs may be in place to direct visitors to another exposure with theropod tracks.

California

Raymond Alf Museum

Claremont, California
Permian, Jurassic, and Tertiary periods
Tracks include *Laoporus* from the Permian of Arizona, theropod tracks from the Jurassic of Utah, and miscellaneous mammal tracks from the Tertiary of California.

The Raymond Alf Museum is part of a remarkable private high school institution, and it is open to the public. It has more fossil footprints on display than does any other institution in the western United States. Half the museum, the lower floor, is devoted to displays of mammal-like reptile, dinosaur, and mammal tracks—mainly from the Permian Coconino Sandstone of Arizona and from the Tertiary (Neogene) of the Mojave desert region.

THE UNIVERSITY OF CALIFORNIA AT BERKELEY

Berkeley, California
Permian, Triassic, Jurassic, Pleistocene eras
Mammal-like reptiles, dinosaurs, and other archosaurs

The University of California at Berkeley has a substantial research collection of tracks from northeastern Arizona, which are not on display. In this way the collection is quite similar to that of the Museum of Northern Arizona and the Raymond Alf Museum because collectors sampled the same areas. Specifically represented are many *Laoporus* specimens from the Permian Coconino Sandstone, *Chirotherium* from the Triassic Moenkopi Formation (including many specimens collected by Frank Peabody), and theropod tracks from the Jurassic Glen Canyon Group. The Collection also contains a replica of the Pleistocene ground sloth tracks from Carson City, Nevada, that were first interpreted as tracks of a giant sandaled human (figure 6.25).

Colorado

DINOSAUR RIDGE

Denver area
Late Jurassic and mid-Cretaceous periods
Mainly ornithopod and theropod dinosaurs

Dinosaur Ridge is a National Natural Landmark situated near the town of Morrison on the western outskirts of the Denver metropolitan area. It was originally famous as the site of the first discovery, in 1877, of the bones of several kinds of dinosaurs—notably, *Stegosaurus* and "Brontosaurus" (now, *Apatosaurus*)—from the late Jurassic Morrison Formation. It has also become famous for the abundant dinosaur tracks of the mid Cretaceous Dakota Group. Previously known as Alameda (meaning "promenade") Parkway, the site has long been used as an outdoor laboratory for earth science students. In recent years there has been much research on the dinosaur tracks, and new tracks have been exposed by the removal of more overburden (see chapter 5). The track beds are part of the so-called Dinosaur Freeway (an enlarged dinosaur promenade). The site has many interpretive signs along a well-designed trail. *A Field Guide to Dinosaur Ridge* is available, as well as a new publication, *Fossil Footprints of the Dinosaur Ridge Area*, and excavated tracks and paleontological exhibits are on display at the nearby Morrison Natural History Center. For information call (303) 697-DINO.

DINOSAUR VALLEY

Grand Junction, Colorado
Mainly Jurassic and Cretaceous periods
Dinosaurs

Dinosaur Valley is the paleontological branch of the Museum of Western Colorado, situated in the heart of Grand Junction. Nominally, it is also the sister institution to Dinosaur Ridge, in eastern Colorado, and the repository for the jointly held University of Colorado and Museum of Western Colorado fossil footprint collection. The museum contains a small dinosaur tracks exhibit (including tracks of ornithopod and theropod dinosaurs and a reconstructed "walkthrough" the coal mine simulating sites from which tracks were excavated. Enthusiastic trackers and dinosaur aficionados should be able to obtain directions to nearby tracksites, such as Cactus Park, and to dinosaur bone sites, which include Rigg's Hill, Dinosaur Hill, and Rabbit Valley. For information call (303) 242–0971 or (303) 241–9210.

Purgatoire Valley

La Junta, in Southeast Colorado
Jurassic period
Dinosaur tracks (sauropod and theropod)

Described as North America's largest dinosaur tracksite, the Purgatoire Valley site is now administered by the U.S. Forest Service, and it is open to the public. This site of Morrison Formation deposits is in a remote canyon that is being preserved in pristine condition; vehicular access is severely restricted. Be prepared to make a day of it. The hike is about five miles each way. Be sure to take water, proper hiking gear, and look out for rattlesnakes and lots of cactus. Mountain bikes and horseback riding are permitted, but check with the Forest Service Office in La Junta for directions to the trail head and for advice. Access is restricted in winter, and overnight camping is not permitted at any time. Explanatory brochures are available at the local Forest Service office. For information call (719) 384–2181.

Rancho del Rio

Statebridge, in Central Colorado
Late Jurassic (Morrison Formation)
Dinosaur tracks (sauropod and theropod)

Rancho Del Rio is a privately owned recreational area and launch site for river rafters. It is situated alongside the Colorado River, about four miles northeast of Statebridge, off Highway 131, between Wolcott and Steamboat Springs. Late Jurassic sauropod tracks from the Morrison Formation (figure 4.36) are located on the opposite side of the river from the Rancho del Rio settlement on public land. They are easily accessible on foot and are well known to local people. No interpretive trails or brochures are provided.

UNIVERSITY OF COLORADO AT DENVER

Denver, Colorado
Carboniferous to Tertiary periods
Diverse tracks
[no specimens on display]

The University of Colorado at Denver, in conjunction with the Museum of Western Colorado, has one of the largest collections of fossil footprints in the western United States. However, the collection is currently used primarily for research and is not on display for the general public. Bona fide researchers are welcome to request access to the collections. The collection forms the basis of the Tracking Dinosaurs exhibit which has toured many venues in Colorado and abroad. For information call (303) 556–2276.

Nebraska

TOADSTOOL PARK, OGALLALLA NATIONAL GRASSLAND

Sioux County, Nebraska
Oligocene Epoch
Tracks of birds and mammals

Tracks from Toadstool Park have only recently been identified, but are clearly very diverse. They range from birds to camels, rhinoceroses and carnivores (chapter 6). For information consult the Ogallalla National Grassland authorities.

New Mexico

CLAYTON LAKE STATE PARK

Near Clayton, in Northeastern New Mexico
Cretaceous period
Dinosaur, Crocodilian

Clayton Lake State Park is an impressive site, revealing a large outcrop of dinosaur tracks in the area of the lake spillway or overflow. The tracksite area has been thoughtfully developed into an interpretive trail, with well-designed walkways and an interpretive pavilion situated on a strategic overlook. The site is wonderfully illuminated late in the day. From a tracker's viewpoint, the site is famous for the controversy about the putative pterosaur tracks, recently shown instead to be of crocodilian affinity. Some reports have claimed that up to eight different track types can be seen here. But the site reveals only tracks of ornithopods (*Caririchnium*), theropods, and crocodilians. In this respect it is typical of other tracksites in the Dakota Group, such as Dinosaur Ridge (see chapter 5).

New Mexico Museum of Natural History and Science

Albuquerque, New Mexico
Permian, Triassic, Jurassic, Cretaceous periods
Dinosaurs and predinosaurian reptiles

The New Mexico Museum of Natural History has recently received the Las Cruces, Permian track collection (Paleozoic Trackway Project, see chapter 2) consisting of diverse trackways of reptiles and amphibians. The collections also contain a variety of Triassic tracks, a few Jurassic footprints and a few replicas of Cretaceous tracks from Clayton Lake State Park. The collection is primarily a research collection but Permian *Dimetropus* tracks and Cretaceous sauropod tracks are on display. For information call (505) 841–8837.

South Dakota

Hot Springs Mammoth site

Pleistocene Epoch
Mammoth tracks

Within the city limits of Hot Springs, South Dakota a large assemblage of mammoth bones has been excavated and turned into an interpretive site accessible to the public. Studies of this site have revealed layers that were extensively trampled by mammoth. Thus the site reveals an interesting combination of footprint and bone evidence.

Texas

Dinosaur Valley State Park

Glen Rose, Southwest of Fort Worth, Texas
Cretaceous period
Dinosaurs (sauropod and theropod)

Dinosaur Valley State Park is the most famous tracksite in Texas. Situated near Glen Rose, on the banks of the Paluxy River, the park offers a small interpretive center and an outdoor exhibit consisting of a replica of the sauropod trackway excavated by the American Museum in New York. The actual outcrops of track-bearing strata in the nearby river bed are usually covered by water and sediment and are therefore hard to see. Any disappointment experienced by the tracker who finds little to see should be mitigated by the knowledge that this site is very significant in the history of dinosaur tracking. One of the largest dinosaur track excavations ever undertaken happened right here. The knowledgeable tracker should ponder the controversy about the "theropod-attacks-sauropod scenario" and ruminate on just what the fascinating trackway evidence (see figure 5.4) really reveals about sauropod social behavior and the hunting or stalking strategies of the large carnivores of the day.

Utah

THE MOAB VICINITY

Moab, Utah
Late Triassic and Jurassic (Chinle, Wingate, Kayenta, Navajo, Entrada-Summerville, and Morrison formations)
Dinosaur tracks and footprints of other archosaurs

Moab is one of the outdoor recreation capitals of the western states. Many new tracksites have been discovered in recent years. It is now possible to visit tracksites in the Navajo Formation, the Entrada-Summerville transition zone, and the Morrison Formation—all within fifteen miles of town. The most easily accessible tracksites are a Navajo Formation tracksite (figure 4.13) alongside the Potash road (Highway 279), seven miles south of the junction with Highway 191, and a Morrison tracksite just off Highway 191, about 23 miles north of Moab, where one can see the trackway of a limping theropod and a turning brontosaur (figure 4.45). Further information on tracksites can be obtained from the Bureau of Land Management (BLM) office in Moab and from the Dan O'Laurie Museum, where some collected tracks from the Kayenta Formation are on display. These institutions also provide explanatory brochures designed by the University of Colorado at Denver Dinosaur Trackers Research Group and by the BLM. The Moab area is proving so rich for dinosaur tracks that the so-called "Dinosaur Triangle" encompassed by Price, Vernal, and Grand Junction may need to be renamed and expanded into a "Dinosaur Diamond," with Moab at the southern apex.

PREHISTORIC MUSEUM, COLLEGE OF EASTERN UTAH

Price, Utah
Cretaceous, Eocene periods
Dinosaur, bird

Situated in the heart of this coal-mining district, the Prehistoric Museum affiliated with the nearby College of Eastern Utah contains the largest collection of Cretaceous dinosaur track casts in the western states. The collection includes various small and large, three- and four-toed footprints (figure 5.31) from the Mesa Verde Group. The museum also contains a few other kinds of tracks, including bird tracks from the Eocene Green River Formation. There are exhibits of skeletal remains of dinosaurs, too, and of other prehistoric animals. Price is part of the so-called "Dinosaur Triangle," bounded by Price, Vernal, and Grand Junction. A small museum at Castle Dale, southwest of Price, has some interesting Eocene unidentified tracks on display.

THE VERNAL VICINITY

Vernal, Utah
Triassic, Jurassic, Cretaceous periods
Dinosaur

Situated at the northern apex of the Dinosaur Triangle, Vernal is in the heart of northeastern Utah's acclaimed dinosaur country. It is near the western end of Dinosaur National Monument, where a famous quarry for dinosaur bones is situated. There are a few tracks in the Dinosaur National Monument collections, but a larger number are on display at the Utah Field House in Vernal. These include a variety of dinosaur om the lower Triassic, the lower and middle Jurassic, and various Cretaceous sites.

WARNER VALLEY

Near Hurricane, Washington County, Utah
Lower Jurassic period
Theropod dinosaur tracks

The Warner Valley dinosaur tracksite, is situated about ten miles southwest of Hurricane, Utah on the south side of Sand Mountain. It can also be reached by traveling southeast from St. George. The site is situated on Bureau of Land Management land (BLM) on a flat sandstone surface in the Moenave Formation. The site exhibits fine examples of large *Eubrontes* tracks, and much smaller *Grallator* tracks, both attributable to theropod dinosaurs. The BLM has provided an interpretive sign at the site, but the interpretation that the *Eubrontes* tracks were made by ten– to twenty–ton prosauropods is incorrect (see chapter 4).

ANNOTATED BIBLIOGRAPHY

Chapter 1: An Introduction to Fossil Footprints

Barthel, K. W., N. H. M. Swinburne, and Conway Morris, S. 1990. *Solenhofen: A Study in Mesozoic Paleontology.* Cambridge, Cambridge University Press, 236 p. (Well illustrated book on the spectacular fossils of the Solenhofen limestones of Germany with illustration of a horseshoe crab dead in its tracks).

Gillette D. D. and M. G. Lockley, eds. 1990. *Dinosaur Tracks and Traces.* New York: Cambridge University Press. 454 pp. (A compilation of papers on dinosaur tracksites from around the world; based on the proceedings of the First International Symposium on Dinosaur Tracks and Traces.)

Halfpenny, J. and E. Biesiot. 1986. *A Field Guide to Mammal Tracking in North America.* Boulder, Colorado: Johnson Books. 161 pp. (One of the best general guides to modern mammal tracks.)

Haubold, H. H. 1984. *Saürierfahrten.* Wittenberg Lutherstadt, Germany: Neue Brehm-Bücherei, A. Ziemsen Publ. 231 pp. (Contains abundant photographs of outstanding specimens.)

Leonardi, G., ed. 1987. *Glossary and Manual of Tetrapod Paleoichnology.* Brasilia, Brasil: Departmento Nacional da Producao Mineral. 117 pp. (Contains a valuable glossary of terms in eight languages, and other interesting information [in English] on fossil tracks and how they have been studied.)

Lockley, M. G. 1986. *Dinosaur Tracksites: A Guide to Dinosaur Tracksites of the Colorado Plateau and American Southwest.* University of Colorado at Denver Geology Dept. Magazine, special issue no. 1., 56 pp. (A field guide published in conjunction with the First International Symposium on Dinosaur Tracks and Traces, Albuquerque, New Mexico; out of print.)

Lockley, M. G. 1986. The paleobiological and paleoenvironmental importance of dinosaur footprints. *Palaois* 1 (1): 37–47. (A brief introduction to the utility of dinosaur tracks.)

Lockley, M. G. 1991. *Tracking Dinosaurs: A New Look at an Ancient World*. Cambridge: Cambridge University Press. 238 pp. (The first popular science book on dinosaur tracks; written for the nonspecialist.)

Lockley, M. G. 1991. The dinosaur footprint renaissance. *Modern Geology* 16: 139–160. (An update of Lockley 1986, with a summary of some of the ideas presented in "Tracking Dinosaurs.")

Lockley, M. G., A. P. Hunt, and C. Meyer. 1994. Vertebrate tracks and the ichnofacies concept: Implications for paleoecology and palichnostratigraphy. In S. Donovan, ed., *The Paleobiology of Trace Fossils*. New York: Wiley & Sons, 241–268. (The first paper to propose and systematically consider the concept of vertebrate ichnofacies.)

Thulborn, R. A. 1990. *Dinosaur Tracks*. London: Chapman and Hall. 410 pp. (A lengthy and expensive text, designed for the specialist. Contains a lot of information, mainly taken from the literature rather than from field studies. Some of this information is therefore outdated and unreliable. Original research by Thulborn on dinosaur gaits and speeds is up to date and interesting.)

Chapter 2: The Paleozoic Era

Alexander, R. M. 1976. Estimates of speeds of dinosaurs. *Nature* 261: 129–130. (Gives formula for calculating speeds of tetrapods from tracks.)

Baird, D. 1952. Revision of the Pennsylvanian and Permian footprints *Limnopus*, *Allopus*, and *Baropus*. *Journal of Paleontology* 26: 832–840. (A good example of a study showing that several named track types are the same.)

Bowlds, L. S. 1989. Tracking the early Permian. *Geotimes* (May): 12–14. (One of the first reports of the tracks discovered in the Abo Formation of the Robledo Mountains.)

Brady, L. F. 1939. Tracks in the Coconino Sandstone, compared with those of small living arthropods. *Plateau* 12: 32–34. (The first in a series of papers [following] on the similarity between modern desert arthropod tracks and those from Permian sand-dune deposits.)

Brady, L. F. 1947. Invertebrate tracks from the Coconino Sandstone of northern Arizona. *Journal of Paleontology* 32: 466–472.

Brady, L. F. 1949. *Onischoidichnus*, a new name for *Isopodichnus* Brady 1947 not Bornemann 1889. *Journal of Paleontology* 23: 573.

Brady, L. F. 1961. A new species of *Paleohelcura* Gilmore from the Permian of northern Arizona. *Journal of Paleontology* 35: 201–202.

Brand, L. R. and T. Tang. 1991. Fossil vertebrate footprints in the Coconino Sandstone (Permian) of northern Arizona: Evidence for underwater origin. *Geology* 19: 1201–1204. (A controversial paper suggesting swimming trackmakers in Coconino time.)

Branson, E. B. and Mehl, M. G. 1932. Footprint records from the Paleozoic and Mesozoic of Missouri, Kansas and Wyoming. *Bulletin of the Geological Society of America*. 43: 383-398. (First report of *Steganosauripus*. Description of new trackways from various localities in the western and southern United States).

Farmer, M. F. 1955. Tracks and trackways of northern Arizona: A record of the past. *Plateau* 28: 54–66. (A record of the location of many tracksites.)

Gilmore, C. W. 1926. Fossil footprints from the Grand Canyon. *Smithsonian Miscellaneous Collections* 77: 1–41. (First in a series of three classic papers describing late Paleozoic tracks from the Grand Canyon area.)

Gilmore, C. W. 1927. Fossil footprints from the Grand Canyon: second contribution. *Smithsonian Miscellaneous Contributions.* 80: 1-80. (Second of three classic papers by Gilmore on the Paleozoic tracks of the Grand Canyon).

Gilmore, C. W. 1928. Fossil footprints from the Grand Canyon: Third contribution. *Smithsonian Miscellaneous Collections* 81: 1–16.

Haubold, H. H. 1984. *Saürierfahrten.* Wittenberg Lutherstadt, Germany: Neue Brehm-Bücherei, A. Ziemsen Publ. 231 pp. (Contains abundant photographs of outstanding specimens.)

Henderson, J. 1924. Footprints in Pennsylvanian sandstones of Colorado. *Journal of Geology* 32: 226–229. (First report of vertebrate tracks from the Paleozoic of Colorado; tracks from the Lyons Sandstone are Permian in age and virtually identical to those from the Coconino.)

Houck, K. and M. G. Lockley, 1986. A field guide to the Pennsylvanian biofacies of the Minturn Formation, Bond-McCoy area, central Colorado trough. *University of Colorado at Denver Geology Department Magazine Special Issue.* 2: 64. (First illustration of tracks from the Belden Formation).

Hunt, A. P., S. Lucas, and P. Huber. 1990. Early Permian footprint fauna from the Sangre de Cristo Formation of northeastern New Mexico. *New Mexico Geological Society Guidebook* 41: 291–303. (First detailed description of tracks from the Sangre de Cristo Formation.)

Hunt, A. P., M. G. Lockley, S. G. Lucas, J. MacDonald, N. Hotton, and J. Kramer. 1993. Early Permian tracksites in the Robledo Mountains, south central New Mexico. *Bulletin of the New Mexico Museum of Natural History and Science* 2: (23–31) (First scientific paper on Robledo tracks.)

Hunt, A. P., M. G. Lockley, and S. G. Lucas, 1994. *Trackway evidence for tetrapod predation from the Early Permian of southern Mexico.* New Mexico Geology, in press. (Brief description of trackway evidence for a pelycosaur attacking a small reptile).

Lockley, M. G. 1989. Middle Pennsylvanian paleoecology of the Eagle Basin region, central Colorado. 21st Congrès International de Stratigraphie et de geologie du Carbonifère, Beijing 1987, *Compte Rendu* 3: 245–250. (First report of tracks from the Minturn Formation.)

Lockley, M.G. 1989. Tracks and traces: new perspectives on dinosaurian behavior, ecology and biogeography. p. 134-145. In K. Padian, and D. J. Chure, (eds). *The Age of Dinosaurs. Short courses in Paleontology #2.* Paleontological Society, Knoxville, Tennessee. (Review of new trackway evidence on dinosaurs).

Lockley, M. G. 1992. Comment: Fossil vertebrate footprints in the Coconino Sandstone. *Geology* 20: 666–667. (Reply to Brand and Tang's controversial 1991 paper; disagrees with the underwater tracks hypothesis.)

Lockley. M. G. and Hunt, A. P. 1994. *Fossil Footprints of the Dinosaur Ridge Area.* A joint publication of the Friends of Dinosaur Ridge and the University of Colorado at Denver. 53 p. (A review of Paleozoic and Mesozoic tracks in the greater Denver area. Contains previously unpublished illustrations of Paleozoic tracks, but also much information on Cretaceous tracks from the immediate Dinosaur Ridge area: see appendix).

Lockley, M.G. and J. Madsen, Jr. 1992. Permian vertebrate trackways from the Cedar Mesa Sandstone of Eastern Utah: Evidence for predator-prey interaction. *Ichnos.* 2:

147–153. (Discussion of probable track evidence for predation by large reptiles on smaller tetrapods.)

Lockley, M. G., A. P. Hunt, and C. Meyer. 1994. Vertebrate tracks and the ichnofacies concept: Implications for paleoecology and palichnostratigraphy. In S. Donovan, ed., *The Paleobiology of Trace Fossils.* New York: Wiley & Sons, pp. 241–268. (The first paper to propose and systematically consider the concept of vertebrate ichnofacies.)

Loope, D. 1992 Comment: Fossil vertebrate footprints in the Coconino Sandstone (Permian) of northern Arizona; evidence for underwater origin. *Geology* 20: 667–668. (Another reply to Brand and Tang 1991; disagrees with the underwater trackmaking hypothesis.)

Lucas, S. G., A. P. Hunt, and M. G. Lockley, 1994. Preliminary report on Permian tracksite, Robledo Mountains, Dona Ana County, New Mexico. *Geological Society of America, Abstracts with Programs.* 26: 27. (Brief overview of main Permian tracksite in southern New Mexico).

Lull, R. S. 1918. Fossil Footprints from the Grand Canyon of the Colorado. *American Journal of Science.* 45: 337-346. (First description of *Laoporus* tracks).

MacDonald, J. P. 1990. *Finding Footprints: Tracking New Mexico's Pre-dinosaurs.* Las Cruces: The Paleozoic Trackways Project. 78 pp. (Preliminary catalog of tracks from the Abo Formation, Robledo Mountains.)

MacDonald, J. P. 1992. Earth's first steps: Footprints from the dawn of time. *Science Probe,* 3:32–47. (Article on Robledo tracks.)

MacDonald, J. P. 1994. *Earth's First Steps.* Johnson Books, Boulder, Colorado (290 pp.). (An autobiographical account of Macdonald's discovery and excavation of Permian footprints in the Robledo Mountains).

McKeever, P. 1991. Trackway preservation in eolian sandstones from the Permian of Scotland. *Geology* 19: 726–729. (Interesting discussion of tracks in various desert sand-dune deposits.)

Spamer, E. E. 1984. Paleontology in the Grand Canyon of Arizona: 125 years of lessons and enigmas from the late Precambrian to the present. *The Mosasaur* 2: 45–128. (Lengthy discussion of Grand Canyon geology, with many notes on tracks by Don Baird.)

Toepelman, W. C. and H. G. Rodeck. 1936. Footprints in late Paleozoic red beds near Boulder, Colorado. *Journal of Paleontology* 10: 660–662. (First report of arthropod trails from the Lyons Sandstone and vertebrate tracks from the Fountain Formation.)

Chapter 3: The Triassic

Branson, E. B. and M. G. Mehl. 1932. Footprint records from the Paleozoic and Mesozoic of Missouri, Kansas, and Wyoming. *Bulletin Geologic Society of America* 43: 383–398. (Named *Agialopous* for a *Grallator*-like track.)

Colbert, E. H. 1989. The Triassic dinosaur *Coelophysis. Museum of Northern Arizona Bulletin,* no. 57, 160 pp. (Monograph on this early theropod.)

Conrad, K. L., M. G. Lockley, and N. K. Prince. 1987. Triassic and Jurassic vertebrate-dominated trace fossil assemblages of the Cimarron Valley region: Implications for paleoecology and biostratigraphy. In S. Lucas and A. P. Hunt, eds., *Northeastern New Mexico: New Mexico Geological Society, 38th Field Conference Guidebook,* pp. 127–138. (Detailed description, including 13 text figures of tracks from the Sloan Canyon and Sheep Pen formations.)

Demathieu, G. R. 1985. Trace fossil assemblages in Middle Triassic marginal marine deposits, eastern border of the Massif Central, France. In A. H. Curran, ed., *Biogenic Structures: Soc. Econ. Paleontol. Mineral. Spec. Publ.* 35: 53–66. (Useful summary in English of typical middle Triassic, European track assemblages.)

Hamilton, A. 1952. The case of the mysterious "Hand Animal." *Natural History* 61: 296–301, 336. (An accessible general reference on *Chirotherium* tracks.)

Hasiotis, S. T. and C. E. Mitchell. 1989. Lungfish burrows in the upper Triassic Chinle and Dolores Formations, Colorado Plateau—Discussion: New evidence suggests origins by a burrowing decapod crustacean. *Journal of Sedimentary Petrology* 59 (5): 871–875. (A convincing argument that burrows previously attributed to lungfish were made by crayfish.)

Haubold, H. H. 1986. Archosaur footprints at the terrestrial Triassic-Jurassic transition. In K. Padian, ed., *The Beginning of the Age of Dinosaurs*, pp. 189–201. New York: Cambridge University Press. (Overview of tracks of this age with proposed track zones or palichnostratigraphy.)

Hunt, A. P. and S. G. Lucas. 1991. *Rioarribasaurus*, a new name for a late Triassic dinosaur from New Mexico. *Palaont. Zeitschr.* 65 (1/2): 191–198. (Renames some specimens of *Coelophysis* as *Rioarribosaurus*.)

Hunt, A. P., V. L. Santucci, M. G. Lockley, and T. J. Olson. 1993. Dicynodont trackways from the Holbrook Member of the Moenkopi Formation (middle Triassic: Anisian), Arizona, USA. *Bulletin of New Mexico Museum of Natural History and Science.* 3:213–218. (First report of therapsid tracks from the Moenkopi Formation.)

Lockley, M. G. 1986. *Dinosaur Tracksites: A Guide to Dinosaur Tracksites of the Colorado Plateau and American Southwest.* University of Colorado at Denver Geology Dept. Magazine, special issue no. 1., 56 pp. (A field guide published in conjunction with the First International Symposium on Dinosaur Tracks and Traces, Albuquerque, New Mexico; out of print.)

Lockley, M. G. and C. Jennings. 1987. Dinosaur tracksites of western Colorado and eastern Utah. In W. R. Averett, ed., *Paleontology and Geology of the Dinosaur Triangle*, pp. 85–90. Grand Junction, Colorado: Grand Junction Geological Society. (First illustration of late Triasic traces from Dolores Valley, western Colorado [162 pp.].)

Lockley, M. G. and A. P. Hunt, 1993. A new Late Triassic tracksite from the Sloan Canyon Formation, type Section, Cimarron Valley, New Mexico. *New Mexico Museum of Natural History and Science Bulletin.* 3: 279-283. (Description of a new tracksite in northeastern New Mexico).

Lockley, M. G., V. F. Santos, and A. P. Hunt, 1993. A new Late Triassic tracksite in the Sheep Pen Sandstone, Sloan Canyon, New Mexico. New Mexico *Museum of Natural History and Science Bulletin.* 3: 285-288. (Details of a new tracksite in the Sloan Canyon Formation that includes new track types).

Lockley, M. G., A. P. Hunt, K. Conrad, and J. Robinson. 1992. Tracking dinosaurs and other extinct animals at Lake Powell. *Park Science: A Resource Management Bulletin* 12: 16–17. (First report of *Atreipus sensu stricto* in the western states.)

Lockley, M. G., K. Conrad, and M. Paquette. 1992. New discoveries of fossil footprints at Dinosaur National Monument. *Park Science: A Resource Management Bulletin* 12: 4–5. (Simple overview of recent late Triassic track discoveries at Dinosaur National Monument.)

Lockley, M. G., K. Conrad, M. Paquette, and J. Farlow. 1992. Distribution and Significance of Mesozoic Vertebrate Trace Fossils in Dinosaur National Monument. *Sixteenth Annual Report of the National Park Service Research Center, University of*

Wyoming, pp. 74–85. (Description of important new Cub Creek site, and first reports of *Pseudotetrasauropus* from western states.)

Lucas, S. G. 1991. Revised Upper Triassic stratigraphy in the San Rafael Swell, Utah. In T. C. Chidsey, Jr., ed., *Geology of East-Central Utah*, pp. 1–8. Utah Geological Association Publication 19. (Reinterpretation of the names of Triassic rocks in Central Utah.)

Lucas, S. G. and A. P. Hunt. 1992. The oldest dinosaurs. *Naturwissenschaften* 79: 171–172. (A discussion of the age of the earliest dinosaurs.)

Olsen, P. E. and D. Baird. 1986. The ichnogenus *Atreipus* and its significance for Triassic biostratigraphy. In K. Padian, ed., *The Beginning of the Age of Dinosaurs*, pp. 61–87. Cambridge: Cambridge University Press. (A detailed study of the ichnogenus *Atreipus*.)

Peabody, F. E. 1947. Current crescents in the Triassic Moenkopi Formation. *Journal of Sedimentary Petrology* 17: 73–76. (An explanation for horseshoe-like markings.)

Peabody, F. E. 1948. Reptile and amphibian trackways from the Lower Triassic Moenkopi Formation of Arizona and Utah. *University of California Bulletin of Geological Science* 28 (8): 295–468. (A classic study of tracks from Moenkopi Formation.)

Chapter 4: The Jurassic

Baird, D. 1980. A prosauropod dinosaur trackway from the Navajo Sandstone (Lower Jurassic) of Arizona. In L. L. Jacobs, ed., *Aspects of Vertebrate History: Essays in Honor of Edwin Harris Colbert*, pp. 219–230. Flagstaff: Museum of Northern Arizona Press. (Description of *Navahopus*.)

Chronic, H. 1983. *Roadside Geology of Arizona*. Missoula, Montana: Mountain Press. 314 pp. (Contains photo of mammaloid tracks that we assign to *Brasilichnium*.)

Colbert, E. H. 1983. *Dinosaurs: An Illustrated History*. New York: Hammond. 224 pp. (Contains photo of Roland Bird's "lost" tracksite near Cameron, Arizona. The site has since been rediscovered.)

Faul, H. and W. A. Roberts. 1951. New fossil footprints from the Navajo(?) Sandstone of Colorado. *Journal of Paleontology* 25 (3): 266–274. (First description of mammaloid tracks from the Navajo Sandstone.)

Haubold, H. H. 1971. Ichnia amphibiorum et reptiliorum fossilium. In O. Kuhn, ed., *Handbuch der Palaoherpetologie, Part 18*. Stuttgart: Gustav Fisher Verlag. 121 pp. (Mammaloid tracks from the Jurassic of South America named as *Tetrapodosaurus*.)

Haubold, H. H. 1984. *Saürierfahrten*. Wittenberg Lutherstadt, Germany: Neue Brehm-Bücherei, A. Ziemsen Publ. 231 pp. (First tabulation of Lower Jurassic track zones.)

Hitchcock, E. 1858. *Ichnology of New England: A Report on the Sandstone of the Connecticut Valley, especially its fossil footmarks*. Natural Sciences of America Reprint. Boston: W. White. 220 pp. (A classic monograph summarizing a life's work on Lower Jurassic tracks from New England. Lull referred to this work as "monumental." It was indeed the beginning of vertebrate ichnology in North America.)

Huene, F. F. von. 1931. Verschedene mosozoishce Wirbeltiereste aust Sudamerika. *N. Jahrb. Min. Geolo. Pal., Abt. B* 66: 181–198. (Mammaloid tracks first reported from the Jurassic of South America.)

Leonardi, G. 1981. Novo icnogenero de tetrapode mesozoico da Formacao Botucatu, Araraquara, SP. *Anais da Academia Brasileira de Ciencias* 53 (4): 793–805. (A thourough study of Jurassic South American mammaloid tracks, and the establishment of the name *Brasilichnium*.)

Lockley, M. G. 1986. *Dinosaur Tracksites: A Guide to Dinosaur Tracksites of the Colorado Plateau and American Southwest*. University of Colorado at Denver Geology Dept. Magazine, special issue no. 1., 56 pp. (A field guide published in conjunction with the First International Symposium on Dinosaur Tracks and Traces, Albuquerque, New Mexico; out of print.)

Lockley, M. G. 1990. Tracking the rise of dinosaurs in eastern Utah. *Canyon Legacy* (Dan O'Laurie Museum in Moab, Utah) 2 (6) :2–8. (First map of an *Otozoum* tracksite from the western states. Includes reference to mammaloid tracks as *Brasilichnium*-like and of probable tritilodont affinity.)

Lockley, M. G. 1991. *Tracking Dinosaurs: A New Look at an Ancient World*. Cambridge: Cambridge University Press. 238 pp. (information on many Jurassic sites).

Lockley, M. G. 1991. The Moab megatracksite: A preliminary description and discussion of millions of middle Jurassic tracks in eastern Utah. In W. R. Averitt, ed., *Guidebook for Dinosaur Quarries and Tracksites Tour, Western Colorado and Eastern Utah*, pp. 59–65. Grand Junction, Colorado: Grand Junction Geological Society. (First and preliminary description of the Moab megatracksite.)

Lockley, M. G., K. Houck, and N. K. Prince. 1986. North America's largest dinosaur tracksite: Implications for Morrison Formation paleoecology. *Geological Society of America Bulletin* 57: 1163–1176. (First description of the Purgatoire dinosaur tracksite; later updated by Lockley and Prince 1988 and Prince and Lockley 1989.)

Lockley, M. G. 1993. *Auf der Spuren der DinoSaürier*. Basel: Birkhaüser. 312 pp. (Translation of *Tracking Dinosaurs*, with additional chapter [no. 11] that introduces new palichnostratigraphic scheme [figure 4.2 herein].)

Lockley, M. G., M. Matsukawa, and A. P. Hunt. 1993. *Tracking Dinosaurs: New Interpretations of Dinosaurs Based on Footprints*. Guidebook to International Dinosaur Exhibit. University of Colorado at Denver. 68 pp. (Includes map of largest mapped site in Moab megatracksite area.)

Lockley, M. G. and N. K. Prince. 1988. The Purgatoire Valley dinosaur tracksite region (Geol. Soc. Am. field guide for centennial meeting). *Colorado School of Mines Professional Contributions* 12: 275–287. (Discussion of the Purgatoire site, including update of track census.)

Lockley, M. G., S. Y. Yang, M. Matsukawa, F. Fleming, and S.-K. Lim. 1992. The track record of Mesozoic birds: Evidence and implications. *Phil. Transactions of the Royal Society London, B* 336: 113–134. (Description of bird-like track *Trisauropodiscus moabensis*.)

Lull, R. W. 1953. Triassic life of the Connecticut Valley. *Connecticut State Geol. Nat. Hist. Surv. Bull.* 81: 1–336. (Classic monograph updating the work of Edward Hitchcock.)

Olsen, P. E. 1980. Fossil great lakes of the Newark supergroup in New Jersey. In W. Manspeizer, ed., *Field Studies of New Jersey Geology and Guide to Field Trips*, pp. 352–398. New York State Geological Assoc., 52nd Annl. Mtg., Rutgers University. (Includes presentation of Olsen's model of the *Grallator-Anchisauripus-Eubrontes* growth series.)

Olsen, P. E., , N. G. MacDonald, Huber and Cornet, B. 1992. Stratigraphy and paleoecology of the Deerfield rift basin (Triassic-Jurassic, Newark Supergroup), Massachusetts. *Guidebook for field trips in the Connecticut Valley region of Massachusetts and adjacent states*. 2: 488-535. (Authors report that purported manus track of *Otozoum* is in fact two small theropod tracks superimposed. Lends support to the idea that the *Otozoum* trackmaker was a biped, probably a prosauropod)

Padian, K. and P. E. Olsen. 1984. The fossil trackway *Pteraichnus*: Not pterosaurian, but crocodilian. *Journal of Paleontology* 58: 178–184.

Prince, N. K. and M. G. Lockley. 1989. The sedimentology of the Purgatoire tracksite region, Morrison Formation of SE Colorado. In D. D. Gillette and M. G. Lockley, eds., *Dinosaur Tracks and Traces*, pp. 155–164. New York: Cambridge University Press. (Updates on the Purgatoire site.)

Stokes, W. L. 1957. Pterodactyl tracks from the Morrison Formation. *Journal of Paleontology* 31: 952–954. (Original description of purported pterosaur tracks.)

Stokes, W. L. 1978. Animal tracks in the Navajo-Nugget Sandstone. *University of Wyoming Contrib. Geol.* 16 (2): 103–107. (Reports of many tracksites in the Navajo Formation.)

Stokes, W. L. and J. H. Madsen, Jr. 1979. The environmental significance of pterosaur tracks in the Navajo Sandstone (Jurassic), Grand County, Utah. *Brigham Young Univ. Geology Studies* 26: 21–26. (More discussion of purported pterosaur tracks.)

Welles, S. P. 1971. Dinosaur footprints from the Kayenta Formation of northern Arizona. *Plateau* 44: 27–38. (Original description of *Hopiichnus*, *Kayentapus*, and *Dilophosauripus*.)

Wellnhofer, P. 1991. *The illustrated encyclopedia of Pterosaurs*. New York: Crescent Books, 192 pp. (Comprehensive and beautifully illustrated volume on all aspects of pterosaurs, including their locomotion).

Chapter 5: The Cretaceous

Bennett, S. C. 1993. Reinterpretation of problematical tracks at Clayton Lake State Park, New Mexico: Not one pterosaur, but several crocodiles. *Ichnos* 2: 37–42.

Bird, R. T. 1941. A dinosaur walks into the museum. *Natural History* 47: 74–81. (An account of the excavation in Texas of the American Museum trackway specimens.)

Bird, R. T. 1944. Did "Brontosaurus" ever walk on land? *Natural History* 53: 61–67. (The publication of the Davenport Ranch tracksite, providing evidence of a herd of brontosaurs.)

Bird, R. T. 1985. *Bones for Barnum Brown: Adventures of a Dinosaur Hunter*, edited by V. T. Schreiber. Fort Worth: Texas Christian University Press. 225 pp. (Roland Bird's posthumous autobiography, and a full account of his work as a dinosaur tracker.)

Brown, B. 1938. The mystery dinosaur. *Natural History* 41: 190–202. (A nonscientific account of the discovery and excavation of hadrosaur tracks from a coal mine in late Cretaceous strata in Colorado; includes the controversial claim of a 15-foot step.)

Currie, P. J., G. Nadon, and M. G. Lockley. 1990. Dinosaur footprints with skin impressions from the Cretaceous of Alberta and Colorado. *Canadian Journal of Earth Sciences* 28: 102–115. (Detailed description of Cretaceous ornithopod tracks with skin impressions.)

Farlow, J. O. (1992). *Sauropod Tracks and Trackmakers: Integrating the Ichnological and Skeletal Records*. Logrono, Spain: Zubia, Inst. de Estudios Riojanos. 10:89–138 (Introduction of the concept of wide and narrow gauge sauropod trackways.)

Farlow, J. O., J. G. Pittman, and J. M. Hawthorne. 1989. *Brontopodus birdi*, Lower Cretaceous sauropod footprints from the U.S. Gulf coastal plain. In D. D. Gillette and M. G. Lockley, *Dinosaur Tracks and Traces*, pp. 371–394. New York: Cambridge University Press. (First formal description of *Brontopodus* tracks.)

Gillette, D.D. and M. G. Lockley. 1989. *Dinosaur Tracks and Traces*. New York: Cambridge University Press. (A fifty-chapter book containing primary sources for information on many of the tracksites and topics discussed in this chapter: ch. 4 deals with R. T. Bird, ch. 7 with purported human tracks, or "man tracks" in Cretaceous beds, ch. 9 with purported brontosaur swim tracks, ch. 15, with Texas tracksites, ch. 34 with the Arkansas tracksite, ch. 39 and ch. 40 with late Cretaceous coal mine tracks.

Horner, J. R. and D. Lessem. 1993. *The Complete T. rex.* New York: Simon and Schuster. 239 pp. (States that *T. Rex* tracks were known at time of publication.)

Lockley, M. G. 1986. *Dinosaur Tracksites: A Guide to Dinosaur Tracksites of the Colorado Plateau and American Southwest.* University of Colorado at Denver Geology Dept. Magazine, special issue no. 1., 56 pp. (A field guide published in conjunction with the First International Symposium on Dinosaur Tracks and Traces, Albuquerque, New Mexico; out of print.)

Lockley, M. G. 1987. Dinosaur footprints from the Dakota Group of eastern Colorado. *Mountain Geologist* 24 (4): 107–122. (First detailed description of Dakota tracks and identification and naming of *Caririchnium* in North America.)

Lockley, M. G. 1987. Dinosaur trackways. In S. J. Czerkas and E. C. Olsen, eds., *Dinosaurs Past and Present*, pp. 80–95. Los Angeles County Museum. (Analysis of Davenport ranch sauropod herd to dispel the myth of adults protecting babies; also see Lockley 1991.)

Lockley, M. G. 1988. Dinosaurs near Denver. *Colorado School of Mines Professional Contributions* 12: 288–299. (First report of Dakota tracks with skin impressions.)

Lockley, M. G. 1990. Did "Brontosaurus" ever swim out to sea? *Ichnos* 1: 81–90. (An argument, and some proof, that sauropod trackways do not indicate swimming sauropods.)

Lockley, M. G. 1990. *A Field Guide to Dinosaur Ridge.* A joint publication of the University of Colorado at Denver and Friends of Dinosaur Ridge. 29 pp. (Includes a description of the Cretaceous tracksite at Dinosaur Ridge, near Denver.)

Lockley, M. G. 1991. *Tracking Dinosaurs: A New Look at an Ancient World.* Cambridge: Cambridge University Press. 238 pp. (A general reference for many Cretaceous tracksites.)

Lockley, M. G. 1994. Dinosaur ontogeny and population structure: Interpretations and speculations based on footprints. In K. Carpenter, K. Hirsch, and J. Horner, eds., *Dinosaur Eggs and Babies.* Cambridge University Press. (Discussion of relationship between track size and age.)

Lockley. M. G. and A. P. Hunt, 1994. *Fossil Footprints of the Dinosaur Ridge Area.* Denver, Friends of Dinosaur Ridge and the University of Colorado at Denver, 53 p. (Popular account of fossil footprints of the greater Denver area, with special emphasis on the Cretaceous).

Lockley, M. G. and A. P. Hunt, 1994. A review of vertebrate ichnofaunas of the Western Interior United States: evidence and implications. in M. V. Caputo, J. A. Peterson, K. J. Franczyk, (eds). *Mesozoic Systems of the Rocky Mountain Region, United States.* p. 95-108. (Synthesis of information on most known Mesozoic tracksites in the western USA).

Lockley, M. G., A. P. Hunt, and C. Meyer. 1994. Vertebrate tracks and the ichnofacies concept: Implications for paleoecology and palichnostratigraphy. In S. Donovan, ed., *The Paleobiology of Trace Fossils.* New York: Wiley & Sons, pp. 241–268. (Defines the *Brontopodus* and *Caririchnium* ichnofacies.)

Lockley, M. G., J. Holbrook, A. P. Hunt, M. Matsukawa, and C. Meyer. 1992. The Dinosaur Freeway: A preliminary report on the Cretaceous megatracksite, Dakota Group, Rocky Mountain front range and high plains, Colorado, Oklahoma, and

New Mexico. In R. Flores, ed., *Mesozoic of the Western Interior Field Trip Guidebook*, pp. 39–54. (Complete description of Dakota tracksites known in 1992, with discussion of the megatracksite, its extent and significance.)

Lockley, M. G. and C. Jennings. 1987. Dinosaur tracksites of western Colorado and eastern Utah. In W. R. Averett, ed., *Paleontology and Geology of the Dinosaur Triangle*, pp. 85–90. Grand Junction, Colorado: Grand Junction Geological Society. (Includes a detailed discussion of tracks from the coal mines of eastern Utah.)

Lockley, M. G., M. Matsukawa, and I. Obata. 1989. Dinosaur tracks and radial cracks: Unusual footprint features. *Bull. Nat. Sci Mus., Tokyo. Ser. C.* 15: 151–160. (Report on discovery of bird and dinosaur tracks together, and excavation of specimen.)

Lockley, M. G., M. Matsukawa, and A. P. Hunt. 1993. *Tracking Dinosaurs: New Interpretations of Dinosaurs Based on Footprints.* Guidebook to International Dinosaur Exhibit. University of Colorado at Denver. 68 pp. (First published map of the Mosquero Creek tracksite, figure 5.20 herein.)

Lockley, M. G. and V. F. Santos. 1993. A preliminary report on sauropod trackways from the Avelino site, Sesimbra region, Upper Jurassic, Portugal. *Gaia: Revista de Geociencias, Museu Nacional de Historia Natural, Lisbon, Portugal* 6: 38–42. (More evidence that manus-dominated sauropod trackways do not indicate swimmers; also discussion of assemblages where sauropod trackway width or gauge is consistent on a given surface.)

Lockley, M. G., B. H. Young, and K. Carpenter. 1983. Hadrosaur locomotion and herding behavior: Evidence from the Mesa Verde Formation, Grand Mesa coalfield, Colorado. *Mountain Geologist* 20: 5–13.

Lockley, M. G., S. Y. Yang, M. Matsukawa, F. Fleming, and S. K. Lim. 1992. The track record of Mesozoic birds: Evidence and implications. *Philosophical Transactions of the Royal Society of London, Ser. B* 336: 113–134.

Mehl, M. G. 1931. Additions to the vertebrate record of the Dakota Sandstone. *American Journal of Science* 21: 441–452. (First report of Cretaceous bird tracks.)

Peterson, W. 1924. Dinosaur tracks in the roofs of coal mines. *Natural History* 24: 388–391. (A short report of purported tyrannosaur tracks.)

Pittman, J. G. 1992. *Stratigraphy and Vertebrate Ichnology of the Glen Rose Formation, Western Gulf Basin, USA.* Ph.D. Thesis. University of Texas at Austin. 726 pp. (The most complete and up-to-date source of information available on the Texas tracksites.)

Robinson, S. F. 1991. Bird and frog tracks from the late Cretaceous of Utah Blackhawk Formation in east-central Utah. *Utah Geological Association Publications* 19: 325–334.

Russell, D. A. and P. Beland. 1976. Running dinosaurs. *Nature* 264: 486. (Discussion of Brown's report as trackway evidence for running duck-billed dinosaurs; but disputed by Thulborn 1981.)

Strevell, C. N. 1932. *Dinosauropes.* Salt Lake City: Deseret News Press (privately published). 15 pp. (An early report on tracks from coal mines in eastern Utah.)

Thulborn, R. A. 1981. Estimated speed of a giant bipedal dinosaur. *Nature* 292: 273–274. (Criticism of the running hadrosaur interpretations of Brown 1938 and Russell and Beland 1976.)

Chapter 6: The Cenozoic Era

Agenbroad, L. D. 1984. Hot Springs, South Dakota: Entrapment and taphonomy of Columbian Mammoth. In P. S. Martin and R. Klein, eds., *Quarternary Extinctions: A Prehistoric Revolution.* Tucson: University of Arizona Press, pp. 113–127. (Formation of the unique assemblage of mammoths and their footprints.)

Alf, R. M. 1966. Mammal trackways from the Barstow Formation, California. *California Academy of Sciences* 65: 258–264. (Description of Miocene tracks from the Mojave Desert.)

Blake, W. P. 1884. The Carson City ichnolites. *Science* 4: 273–276. (Early report on the Nevada mammal tracks.)

Brady, L. F. and P. Seff. 1959. "Elephant Hill." *Plateau* 31: 80–82. (Reports Mammoth tracks from the vicinity of Montezuma Castle National Monument, Arizona.)

Breed, W. J. 1973. New Avian Fossils from the Bidahochi Formation (Pliocene), Arizona. *In* Fassett, J. E. (ed.) Cretaceous and Tertiary Rocks of the southern Colorado Plateau. *Four Corners Geological Society Guidebook* pp. 144–147. (Description of goose tracks similar to those of modern Canada geese).

Chaffee, R. G. 1943. Mammal footprints from the White River Oligocene. *Natulae Naturae*. Philadelphia: Academy of Natural Sciences of Philadelphia, 116: 1–13. (Mammal tracks from the bone-rich White River Group.)

Curry, D. H. 1957. Fossil tracks of Eocene vertebrates, southwestern Unita Basin, Utah. In Intermountain Association of Geologists, *Eighth Annual Field Conference Guidebook*, pp. 42–47. (Tracks from the Green River lake deposits.)

Erickson, B. R. 1967. Fossil bird tracks from Utah. *Museum Observer* 5: 6–12. (Eocene bird tracks from the Green River lake deposits.)

Gilmore, C. W. 1928. Fossil footprints from the Fort Union (Paleocene) of Montana. *Proceedings of the United States National Museum* 74: 1–4. (Account of rare Paleocene tracks.)

Greben, R. and M. G. Lockley. 1992. Vertebrate tracks from the Green River Formation, eastern Utah: Implications for paleoecology. *Geological Society of America, Abstracts with Programs* 24: 16. (Overview of the Eocene Green River tracks.)

Johnson, K. R. 1986. Paleocene bird and amphibian tracks from the Fort Union Formation, Bighorn Basin, Wyoming. *Contributions to Geology, University of Wyoming* 24:1–10. (Further description of Paleocene tracks.)

Laury, R. L. 1980. Paleoenvironment of a Late Quaternary Mammoth-bearing Sinkhole Deposit, Hot Springs, South Dakota. *Bulletin of the Biological Sciences of America.* 91: 465–475. (Describes tracks associated with mammoth skeletons.)

Lozinsky, R. L. and Tedford, R. H. 1991. Geology and paleontology of the Santa Fe Group, southwestern Albuquerque basin, Valencia County, New Mexico. *New Mexico Bureau of Mines and Mineral Resources Bulletin.* 132: 1-30. (Geologic report that includes photographs and a drawing of a camel trackway).

Marche, J. D. II. 1986. Extraordinary petrifactions: Footprints at Nevada State Prison. *Terra* 24 (5): 12–18. (Account of the discovery of the Carson City tracks.)

Martin, L. D. and D. K. Bennett. 1977. The burrows of the Miocene beaver *Paleocastor*, western Nebraska, USA. *Palaeogeography, Palaeoclimatology, Palaeoecology* 22: 173–193. (Explanation of mysterious large burrows.)

Nations, J. D., D. W. Hevly, D. W. Blinn, and J. J. Landye. 1981. Paleontology, paleoecology, and depositional history of the Miocene-Pliocene Verde Formation, Yauapai County, Arizona. *Arizona Geological Society Digest* 13: 133–149. (Includes discussion of the Verde tracks.)

Nininger, H. H. 1941. Hunting prehistoric lion tracks in Arizona. *Plateau* 14: 21–27. (Discovery of lion tracks in Arizona.)

Nixon, D. A. and LaGarry-Guyon, H. E. 1993. New trackway site in the White River Group type section at Toadstool Park, Nebraska: paleoecology of an Oligocene braided stream, riparian woodland, and adjacent grassland. *Journal of Vertebrate Paleontology.* 13 (supplement to number 3): 50A. (Preliminary report on very diverse vertebrate trackways from a variety of environments that indicate much paleoecological information such as which animals were solitary and which travelled in herds; see appendix).

Packard, E. L. and Allison, I. S.. 1980. Fossil Bear Tracks in Lake County, Oregon. *Oregon Geology* 42: 1–2. (Description of tracks made by a huge extinct bear.)

Peabody, F. 1954. Trackways of an ambystomid salamander from the Paleocene of Montana. *Journal of Paleontology* 28: 79–83. (A careful identification of rare amphibian tracks.)

Sargeant, W. A. S. and J. A. Wilson. 1988. Late Eocene (Duchesnean) mammal footprints from the Skyline Channels of Trans-Pecos Texas. *Texas Journal of Science* 40: 439–446. (Rare Eocene tracks which are not from the Green River.)

Sargeant, W. A. S. and W. L. Langston, Jr. 1994. Vertebrate footprints and invertebrate traces from the Chadronian (late Eocene) of Trans-Pecos Texas. *Bulletin of Texas Memorial Museum.* 36: 1-86. (Detailed description of the most diverse trackway assemblage in the Tertiary of North America).

Scrivner, P. J. and D. J. Bottjer. 1986. Neogene avian and mammalian tracks from Death Valley National Monument, California: Their context, classification, and preservation. *Palaeogeography, Palaeoclimatology, Palaeoecology* 57: 285–331. (Detailed study of diverse tracks from Death Valley.)

Webb, S. D. 1972. Locomotor evolution in camels. *Forma et Functio.* 5: 99-112. (Describes bone and trackway evidence for the origin of the distinctive pacing gait of camels and llamas).

Chapter 7: Tracks Galore

Chure, D. 1992. Biodiversity—it's more than just biological. *Park Science: A Resource Management Bulletin* 12: 7. (A valuable short comment on the fact that fossils are part of the earth's total biodiversity record.)

Lockley, M. G. 1989. Tracks and traces: New perspectives on dinosaurian behavior, ecology, and biogeography. In K. Padian and D. J. Chure, eds., *The Age of Dinosaurs*, Short Courses in Paleontology, No. 2. Knoxville, Tenn.: Paleontological Society. (The first paper proposing a classification of formations according to relative abundance of bones and tracks.)

Lockley, M. G. 1991. *Tracking Dinosaurs: A New Look at an Ancient World*. Cambridge: Cambridge University Press. 238 pp. (Contains a chapter on using tracks as a tool for paleoecological census, and a classification of formations according to track content, following Lockley 1989; also contains a chapter on trampling.)

Lockley, M. G. 1992. La dinturbacion y el fenomeno de la alteracion del sedimento por pisadas de vertebrados en ambientes antiguos. In J. L. Sanz, and A. D. Buscalioni, eds., *Los Dinosaurios y su Entorno Biotico*, Actas del Segundo de Paleontologia en Cuenca, pp. 272–296. Cuenca, Spain: Instituto Juan de Valdes. (The first paper dealing specifically with the phenomenon of vertebrate trampling and dinoturbation; also see Lockley 1991.)

Lockley M. G. The paleoecological and paleoenvironmental utility of dinosaur tracks in J. O. Farlow, and M. Brett-Surman, (eds.) *Dinosaurs: A Sesquicentennial Celebration*. University of Indiana Press, in press. (First attempt to use track census data to estimate biomass in a particular dinosaur community).

Lockley, M. G. and A. P. Hunt. 1993. Fossil footprints: A previously overlooked paleontological resource in Utah's national parks. In V. L. Santucci, ed., *National Park Service Paleontological Research Abstract Volume*, p. 29. Denver: U.S. Department of Interior, Natural Resources Publications Office. (Contains a comparison of the track and bone records at Dinosaur National Monument, and an estimate of the extent of the track record at other sites.)

Lockley, M. G. and A. P. Hunt. 1994. A review of vertebrate ichnofaunas of the western interior United States: Evidence and implications. In M. V. Caputo, J. A. Peterson, and K. Franczyk, eds., *Mesozoic Systems of the Rocky Mountain Region, United States.* pp. 95–107. (The first paper presenting the classification of formations according to their track and bone content.)

Lockley, M. G., A. P. Hunt, and C. Meyer. 1994. Vertebrate tracks and the ichnofacies concept: Implications for paleoecology and palichnostratigraphy. In S. Donovan, ed., *The Paleobiology of Trace Fossils.* New York: Wiley & Sons, pp. 241–268. (Defines vertebrate ichnofacies, and presents track census data for the Chinle Group and various Cretaceous deposits.)

Lockley, M. G., J. Holbrook, A. P. Hunt, M. Matsukawa, and C. Meyer. 1992. The Dinosaur Freeway: A preliminary report on the Cretaceous megatracksite, Dakota Group, Rocky Mountain front range and high plains, Colorado, Oklahoma, and New Mexico. In R. Flores, ed., *Mesozoic of the Western Interior Field Trip Guidebook.* Fort Collins, Colorado, pp. 39–54.

Appendix: Fossil Footprint Sites

Barnes, F. A. 1986. *Utah Canyon Country.* Utah Geographic Series, Inc. Salt Lake City, 120 p. (Includes illustration of theropod tracks, from Explorer Canyon, that are similar to those on display at the visitors center in Page).

Brand, L. 1997. *Faith, Reason and Earth History.* Andrews University Press.

Davies, D. 1992. *In Pictures Glen Canyon-Lake Powell-the Continuing Story.* KC Publications Inc. Las Vegas, Nevada, 48 p. (Nice illustration of *Otozoum* tracks on page 42).

Ellenberger, P. 1972. Contribution a la classification des piste de vertebres du Trias: les types du Stormberg d'Afrique du Sud (1). *Palaeovertebrata,* Memoire Extraordinaire, Laboratoire de Paleontologie des Vertebres, Montpellier.

Foster, J. R., and Lockley, M. G. 1997. Probable crocodilian tracks and traces from the Morrison Formation (Upper Jurassic) of eastern Utah. *Ichnos* 5:121–129.

Hunt, A. P., and Lucas, S. G. 1998. Ichnological evidence for predation in the Paleozoic. Is there any? In Lucas, S. G. et al. (eds) *New Mexico Museum of Natural History and Science Bulletin* 12, pp. 59–62.

I.C.Z.N. 1996. Opinion 1942. Coelurus barui Cope, 1887 (currently Coelophysis bauri; Reptilia, Saurischia): lectotype replaced by a neotype. *Bulletin of Zoological Nomenclature* 53:142–144.

Kelsey, M. R. 1989. *Boater's Guide to Lake Powell.* Kelsey Publishing, Provo Utah. 288 p. (Illustration of an directions to Wingate Formation tracksite at the Rincon).

Lee, Y-N. 1997. Bird and dinosaur footprints in the Woodbine Formation (cenomanian), Texas. *Cretacious Research* 18:849–864.

Leonardi, G. and Lockley, M. G. 1995. A proposal to abandon the ichnogenus *Coelurosaurichnus* Huene, 1941—a junior synonym of *Grallator* E. Hitchcock 1858. *Journal of Vertebrate Paleontology* 15:40A.

Lockley, M. G., Hunt, A. P., Gaston, R., and Kirkland, J. 1996. A trackway bonanza with mammal footprints from the late Triassic of Colorado. *Society of Vertebrate Paleontology Bulletin* 16:49A.

Lockley, M. G., Hunt, A. P., Paquette, M., Bilbey, S. A., and Hamblin, A. 1998. Dinosaur tracks from the Carmel Formation, northeastern Utah: implications for Middle Jurassic paleoecology. *Ichnos* 5:255–267.

Lockley, M. G., Meyer, C. A., and dos Santos, V. F. 1996. *Megalosauripus, Megalosauropus* and the concept of Megalosaur footprints. In Morales, M. (ed.) *Continental Jurassic Symposium Volume,* Museum of Northern Arizona. pp. 113–118.

Miller W. E., B. B. Britt, and K. L. Stadtman. 1989. Tridactyl trackways from the Moenave Formation of Southwestern Utah. *In* D. D. Gillette, and M. G. Lockley, (eds.). *Dinosaur Tracks and Traces.* Cambridge University Press. 454 pp. (A description and map of the Warner Valley tracksite. Paper includes discussion of authors who have attributed *Eubrontes* tracks to theropods rather than to prosauropods).

GENERAL INDEX

SYSTEMATIC ICHNOLOGICAL INDEX